Network-on-Chip - Architecture, Optimization, and Design Explorations

Edited by Isiaka A. Alimi,
Oluyomi Aboderin, Nelson J. Muga
and António L. Teixeira

Published in London, United Kingdom

IntechOpen

Supporting open minds since 2005

Network-on-Chip - Architecture, Optimization, and Design Explorations
http://dx.doi.org/10.5772/intechopen.91110
Edited by Isiaka A. Alimi, Oluyomi Aboderin, Nelson J. Muga and António L. Teixeira

Contributors
Masaru Fukushi, Yota Kurokawa, Erulappan Sakthivel, Rengaraj Madavan, Adebayo E. Abejide, Madhava R. Kota, Sushma Pandey, Oluyomi Aboderin, Mário Lima, António L. Teixeira, Cátia Pinho, Isiaka Ajewale Alimi, Romil K. Patel, Abdelgader M. Abdalla, Nelson J. Muga, Armando N. Pinto, Ramoni A. Gbadamosi, Zhongjing Ren, Jianping Yuan, Peng Yan

First published in London, United Kingdom, 2022 by IntechOpen
IntechOpen is the global imprint of INTECHOPEN LIMITED, registered in England and Wales, registration number: 11086078, 5 Princes Gate Court, London, SW7 2QJ, United Kingdom
Printed in Croatia

British Library Cataloguing-in-Publication Data
A catalogue record for this book is available from the British Library

Additional hard and PDF copies can be obtained from orders@intechopen.com

Network-on-Chip - Architecture, Optimization, and Design Explorations
Edited by Isiaka A. Alimi, Oluyomi Aboderin, Nelson J. Muga and António L. Teixeira
p. cm.
Print ISBN 978-1-83968-148-6
Online ISBN 978-1-83968-158-5
eBook (PDF) ISBN 978-1-83968-159-2

For EU product safety concerns:
IN TECH d.o.o., Prolaz Marije Krucifikse Kozulić 3, 51000 Rijeka, Croatia, info@intechopen.com or visit our website at intechopen.com.

Meet the editors

Isiaka A. Alimi received his Ph.D. in Telecommunications Engineering from the University of Aveiro, Portugal. He was with the Federal Radio Corporation of Nigeria as a senior engineer (RF transmission and management) and the Department of Electrical and Electronics Engineering, Federal University of Technology, Akure, Nigeria, as a lecturer. He is currently a researcher at the Instituto de Telecomunicações, Aveiro, Portugal, where he has been participating in various R&D activities. He has authored/co-authored more than forty technical papers, nine book chapters, and has co-edited one book. His research interests include optical communications, microwave photonics, network security, fixed-mobile broadband (wired and wireless technologies) convergence, and their applications for effective resources management in access networks. He is a member of the Institute of Electrical and Electronics Engineers (IEEE).

Oluyomi Aboderin obtained a Ph.D. in Telecommunications Engineering from the University of Porto, Portugal. He also obtained a master's degree in Personal Mobile and Satellite Communications from the University of Bradford, United Kingdom, and a bachelor's degree from the Ladoke Akintola University of Technology, Nigeria. He joined the National Space Research and Development Agency in 2005 and is currently an assistant director with the agency, in the frequency coordination and management team. He was a researcher with Instituto de Telecommunicações, Aveiro, Portugal, and has been participating in various research and development activities. He has published more than ten technical papers, including a patent. His research interests include satellite channel modeling, frequency management, antenna design, underwater communications, and microwave photonics.

Nelson Muga graduated in Physics from the University of Porto, Portugal, in 2002. He received a master's degree in Applied Physics in 2006, and a Ph.D. in Physical Engineering in 2011, both from the University of Aveiro, Portugal. He has been a lecturer in the Physics Department, the University of Aveiro since 2016, where he teaches courses related to optics, optoelectronics, and photonics. Currently, he is an auxiliary researcher at the Instituto de Telecomunicações, Aveiro, where, over the years, he has participated as a researcher in more than twenty-five R&D projects, developing expertise in the field of high-speed optical communications and quantum-secure optical communication systems and technologies. He has published more than forty papers in international journals and more than sixty international conference proceedings. He is a senior member of the Optical Society.

Antonio Teixeira obtained a Ph.D., partly developed at the University of Rochester, USA, from the University of Aveiro, Portugal, in 1999. He holds an Executive Certificate in Management and Leadership from the MIT Sloan School of Management, Massachusetts, USA, and a post-graduate degree in Quality Management in Higher Education. He joined the University of Aveiro in 1999 and is presently a full professor and a senior researcher in the Instituto de Telecomunicações. Since 2014, he has been the dean of the Doctoral School, University of Aveiro. Dr. Teixeira has worked at several industrial organizations, including Nokia Siemens Networks, Coriant, and PICadvanced, a startup in photonics that he cofounded that employs more than forty highly skilled persons. He holds eleven patents and has published more than 400 papers. He has supervised more than seventy MSc and fifteen Ph.D. students and has participated in more than thirty-five national and international projects.

Contents

Preface

On-chip communication has been experiencing unprecedented pressure due to the huge amount of intellectual property (IP) cores that can now be integrated on a single chip. For traditional bus-based interconnections, the integration results in scalability and contention issues for on-chip communication. Based on this, they are limited and unable to effectively support the required inter-component communication in the System-on-Chip (SoC). Consequently, the major challenges in many core-based SoCs are related to scalability, flexibility, and a high-performance communication backbone.

Network-on-Chip (NoC) has emerged as an efficient solution for offering the required architectural flexibility and parallelism to support the associated massive cores and IPs. In an NoC system, communication between the cores is a router-based packet-switched transmission. Based on this, for an optimum trade-off between flexibility and energy efficiency, there has been an increase in the implementation of NoC architectures for on-chip communications in embedded multicore processors such as Multiprocessor SoCs (MPSoCs), Chip Multiprocessors (CMPs), and Graphics Processing Units (GPUs). In addition, multicore processing is attractive for power reduction in general-purpose computing and embedded systems.

This book covers the fundamental concepts and the state of the art of NoC architecture. This comprises the exploration of process and component optimization. It also focuses on the cost-effective and appropriate combination of components and processes. In this context, hybrid modulation can be employed for Photonic Integrated Circuits (PICs) to ensure high-performance communication. A traffic-aware sense amplifier can also be employed in an NoC system to alleviate energy consumption. Furthermore, the book focuses on various network architecture and designs for NoC systems. In this regard, recent advances in design-friendly, scalable, flexible, and high-performance interconnection architectures are presented along with the associated technical challenges and research direction for design optimization. Moreover, microstructure fabrication and routing optimization are also covered in this book, as the employed routing algorithm can significantly influence the overall network performance metrics regarding latency and throughput.

In general, this book not only presents underlying concepts, features, and related evolutions but also clarifies the fundamental technical principles of on-chip communications with good insights into future NoC systems. The information presented is easy to follow, concise, and comprehensible. It comprises both theoretical and practical areas of system implementation. This makes it suitable for students, researchers, and professional engineers. It is also a good reference for all interested readers who wish to keep abreast of the current trends in on-chip communications, especially NoC systems.

The editors would like to acknowledge and appreciate all the contributing authors. Their works and innovations in different areas of NoC systems are highly appreciated. The editors would also like to thank IntechOpen for the invitation to participate in this project.

Isiaka A. Alimi, Oluyomi Aboderin and Nelson J. Muga
Instituto de Telecomunicações,
Aveiro, Portugal

António L. Teixeira
Department of Electronics,
Telecommunications and Informatics,
University of Aveiro and the Instituto de Telecomunicações,
Aveiro, Portugal

Section 1

Process and Component Optimization

Chapter 1

Direct and External Hybrid Modulation Approaches for Access Networks

Adebayo E. Abejide, Madhava R. Kota, Sushma Pandey, Oluyomi Aboderin, Cátia Pinho, Mário Lima and António Teixeira

Abstract

The demand for low-cost high-speed transmission is a major challenge for 5G future networks. To meet this optical communication demand, holistic and pains-taking approaches are required in designing a simplified system model. Since the demands for high bandwidth are growing at unprecedented speed as we approach the Zettabyte era, it is crucial to minimize chromatic dispersion (CD) associated to high bit-rate signals. Mitigating CD electronically comes at high cost which may not be compatible with 5G. Photonic Integrated Circuit (PIC) as an enabler for fast speed optical transmission is still undergoing its growth stage and its major speed and efficiency have not yet been attained. However, proper and right combination of components and approaches can potentiate this technology in a more cost-efficient way. Hybrid modulation (HM)-PIC presents a simplified approach in terms of cost and efficiency for 5G networks. Hybridization of existing modulation components and approaches in PIC can enhance the generation of high bit-rate signals without the need for electrical CD compensation. A detailed study of hybrid multilevel signal modulation concept as a valuable solution for Data Centers (DC) high data-rate signals and next-generation Passive Optical Networks (PONs) is proposed.

Keywords: photonic integrated circuits (PIC), hybrid modulation (HM), direct modulation laser (DML), external modulation laser (EML), simplified high bit-rate signal generation, chromatic dispersion (CD), insertion loss (IL), chirp managed lasers (CML), optical spectrum reshaper (OSR)

1. Introduction

In the era of explosive demand for bandwidth and complex broadband transmission attributed to the fifth-generation (5G) and beyond, strategic positioning of fiber optics communications is imperative considering its huge advantages in terms of high-speed transmission, simplified design and low cost implementation [1].

Fiber Optics communication, which is carried out through photon, has been a promising approach that can conveniently guarantee the requirements of high data transmission. However, the existing technologies will require complete overhauling, upgrading, or redesigning for effective results [2]. Unarguably the 5th and 6th generations of wireless networks cannot sufficiently guarantee their expected output in terms of speed, latency and spectral efficiency without improved channel architectures.

Therefore, there is an urgent need to further optimize optical properties to meet the current and future demands for high-speed data communications [3]. Fiber optics on its own with peak-to-peak throughput of 100 Gb/s is unique among other communication channels, and as of today, its capacity in terms of speed and convenience of communication has not been fully utilized [1].

Similarly, PIC as an enabler for optical communications presents a promising approach for conveniences of communications in terms of low footprint, low cost implementation, and very high speed [2, 4]. Therefore, increasing data rate and mobility in the evolving technologies require an increase of data traveling in the networks, introducing new requirements (e.g., speed and latency) for the components, e.g., PIC building blocks (BB). Although, PIC components under terahertz speed is a promising technique to guarantee the requirements of future networks, the maturation of the integrated photonics is still ongoing.

Furthermore, to strengthen the position of fiber optical communications for the emergence of hyperscale future transmission such as hyperscale data centers (HDCs) and to continually improve the amount of data transmission over optical networks through modulation and multiplexing approaches, this chapter will address a detailed description of DML and EML modulation schemes, and improved hybrid approaches, a valuable solution to attain the overcoming requirements of high-speed signal transmissions.

Conventional approaches are carried out through direct modulation laser (DML) and external modulation laser (EML) respectively which cannot guarantee the demands of future networks due to their respective limitations [2, 5, 6]. Although DML is simple in design and can generate high power budget when properly optimized but, the signal is degraded due to high CD, induced adiabatic chirp, phase noise, and low extinction ratio (ER). Consequently, these limit DML transmissions to a short distance of about 10 km and its inability to cope with high bit-rate transmissions [5–7].

EML on the order hand outperforms DML with reduced chirp and better reach. Nevertheless, the signal is also limited due to size, low optical power and high driving voltage among others [2, 8, 9]. Hence, a hybrid combination and optimization of these modulation processes can be seen as a powerful approach.

With the HM method, high bit-rate signals can be generated in a simplified way with a low footprint, low energy consumption and less complexity in meeting the demand for future networks [2].

This chapter provides a detailed study of the development and optimization of a simplified optical transceiver for high data rate 5G optical transmission, in order to manage future competitive markets necessities in its migration to 100G and 400G Ethernet, by replacing the single-mode 10G-SFP+ used for 4G networks. Moreover, the access and aggregation layers patterned with Dense Wavelength Division Multiplexing (DWDM) transceivers having bit-rate as high as 400G designed for metro access, metro convergence and core layer access networks, are also studied.

With HM approach, simplified and high bit-rate intensity modulated signals can be generated without electrical pulse shaping, digital to analog conversion (DAC), and digital signal processing (DSP) compensation [2].

The remaining part of this chapter is arranged as follows: in section 2, details signal modulation approach for PIC is discussed follow by section 3 where we presented the novel HM approach and the results of our simulations. We draw the conclusion in section 4.

2. Signal modulation in PIC

The crucial function of modulators in optical communications is the conversion of the electrical information input signal into its corresponding optical domain which is placed on the optical signal as a carrier before being launched into an optical communications channel. This process is carried out in the optical transmitter, where an optical signal from an optical source such as a semiconductor laser or light-emitting diode (LED) is either directly or externally modulated. Optical signal has three major properties which are amplitude, frequency and phase. However, the message electrical signal is biased to manipulate any of these optical properties so that information can be sent along optical channels [2, 9].

Modulation of the optical carrier properties mentioned above could be done directly or externally [5]. Direct modulation is achieved using a DML [8], while the external modulation, which is equally referred to as external modulation laser (EML), can be either through electro-refractive using Mach Zehnder Interferometer (MZI) [9] or electro-absorption using an electro-absorption modulator (EAM) [7, 9].

The simplicity of design and cost [9, 10] make DML a preferred choice, nonetheless several constraints make it undesirable for high bitrate and long-distance transmission [7]. These constraints include low bandwidth, low efficiency, CD caused by induced chirp which imposes signal phase noise, refractive index change of the active layer by carrier density modulation and relatively low ER [5, 6]. DML induced chirp can be transient or adiabatic. Chirp due to transient gives a nonlinear gain which occurs during the bit transitions, while the adiabatic chirp is the spontaneous emission that occurs in the laser, which is responsible for the blue-shifting of bit 1 relative to bit 0 [11]. As a result, the increase of bit-rate in DML causes the signal to suffer a very pronounced pulse broadening due to CD [6], mainly when the bit-rate is increased beyond 10 Gb/s [12]. The chirp in DML is mainly influenced by the linewidth enhancement factor, which is also known as Henry factor limiting within 2 and 8 for DFB laser, and consequently, the signal of 10 Gb/s transmission could not go beyond 10 km [13].

On the other hand, external modulation approaches can achieve high ER, high data-rate, and lower modulation distortion. Unfortunately, this comes with additional system complexity and cost compared to direct modulation [14]. For the external modulation, EAM offers lower cost and size combining with a considerably higher speed when compared to MZI [9, 15]. EAM can achieve better speed with low CD as a result of its zero or negative chirp [6] with the inherent advantage of low driving voltage and the possibility of monolithic integration with a DFB laser on a single waveguide, which reduces channel insertion loss (IL) [9, 16, 17]. Furthermore, EAM optical modulation does not affect the laser properties unlike DML [7] and its low CD can achieve improved speed (25–40 Gb/s) and longer distance (10–40 km) [7, 9]. Additional details regarding the two technologies are provided in the subsections 2.1 and 2.2. Presented results were attained by VPIphotonics® transmission simulations.

2.1 DML signal generation

In this approach, an electrical signal is directly injected into the lasing cavity in an attempt to manipulate the stimulated emission present in the laser cavity. By this, a high-frequency electromagnetic signal with information could be sent via optical channel after modulation [9, 14]. **Figure 1** gives an illustration of how the DML approach can be achieved and its chirp effects.

Practically, semiconductor lasers such as distributed feedback laser (DFB), constricted-mesa lasers and Fabry-Perot (FP) lasers are the major lasers used for

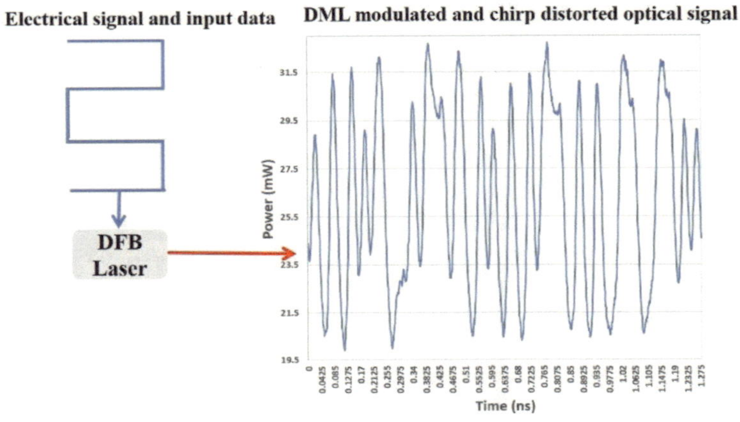

Figure 1.
DML showing electrical signals without chirp before modulation and optical signal after modulation with chirping effect.

DML purposes. It is important to note that DML is simple and cost efficient but can only be applied with low bit-rate and short-reach as a result of chirp due to spectral broadening associated with the biasing current during modulation. This is due to accompanying phase modulation (PM) to the desired intensity modulation (IM) during this process [18]. The level of the chirping imposed on the modulated pulse largely depends on the driving condition (I_{bias} and modulation current (ΔI)) and the laser type.

2.1.1 DML chirping effects

The chirp associated with DML can be subdivided into transient and adiabatic chirp as depicted in **Figure 2**. The presence of chirp in the modulated pulse leads to high CD and invariably inter-symbol-interference (ISI) which adversely weakens the reach and effects of DML transmission.

According to [19], transient chirp is associated with the relatively small frequency difference between the steady-state of signal pulse levels of ones and zeros. This frequency difference leads to ringing and significantly obvious overshooting of optical output power and frequency deviations.

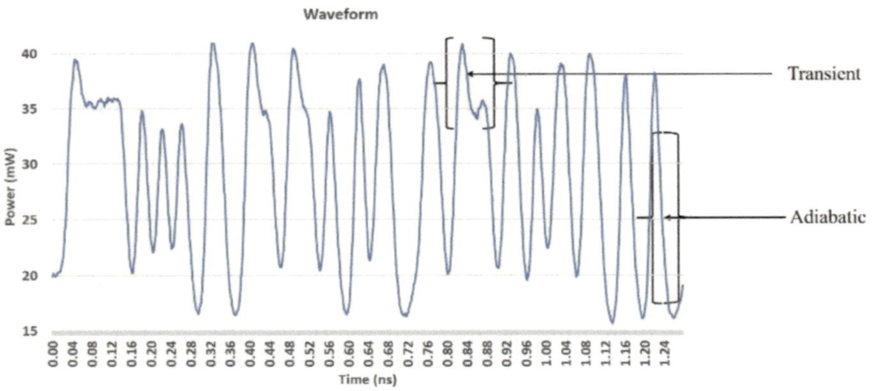

Figure 2.
DML power time domain response showing transient and adiabatic chirp.

Adiabatic chirp at the same time can be described with the damping oscillations and large frequency differences between pulse ones and zeros.

Compensating the chirp and CD in DML comes with its own cost and complexity. Some of these approaches are the use of dispersion compensation fiber (DCF), electrical signal compensations through decision feedback equalizer (DFE), feed-forward equalizer (FFE), continuous-time linear equalizer (CTLE), digital signal processing (DSP), electrical pulse shaping and finally, optical CD compensation through optical spectrum reshape (OSR) [20]. This OSR method is better detailed in the subsection 3.3.

Nevertheless, progressive works to reduce the effect of chirp on DML show improved results. For instance, in [21], an InGaA1As/InGaA1As multi-quantum well (MQW)-DML was grown on an n-doped InP substrate by shortening the laser cavity length to less than 150 μm via positioning of a passive waveguide in the front of the DFB laser. The DML was operated at 45 mA driving current with 43 Gb/s bit-rate, the optical signal obtained after 40 km presents a clear eye signal at 25 °C which shows an improved bit-rate possibility higher than 10 Gb/s over DML.

Similarly, the research work in [22] also presents a 40 Gb/s DML optical signal that used passive feedback laser (PFL) realized around 1300 nm and 1550 nm wavelength region. In this work, DFB and integrated passive feedback section (IPF) were combined enabling the suppression and control of the phase feedback field with the modulation performance of the stationary operating laser.

Other works in [23–25] also present similar improvements in the design of DML with higher bit-rate and improved fast speed transmission. The authors in [23] used InP-on-Si to achieve 45 Gb/s and 25 GHz 3-dB modulation bandwidth at a relaxation oscillation frequency of 10 GHz using NRZ encoder. The fabricated laser has two sections of around 250 μm each with an active region on InGaAsP separate confinement hetero-structure (SCH). A clear eye diagram after a 2 km transmission of 45 Gb/s was obtained with BER lower than 7% Hard Decision (HD) forward error correction (FEC) and received optical power (ROP) around −7 dBm.

2.1.2 DML signal optimization

To further investigate the behavior of DML, we carried out a simulation study of a rate equation model of a semiconductor DFB laser. In order to reduce the effect of the unwanted transient chirp on the modulated signal, it is highly advised to bias the laser far away from the current threshold [9, 26], allowing to obtain high optical output power after modulation. Although, this is a trade-off with the optical signal ER. In our case, the laser current threshold is obtained around 10 mA and as we biased away from this point, the output optical signal increases with reduced transient chirp while the signal ER reduces.

The block diagram for an intensity DML is presented in **Figure 3** and the DML parameters used are summarized in **Table 1**.

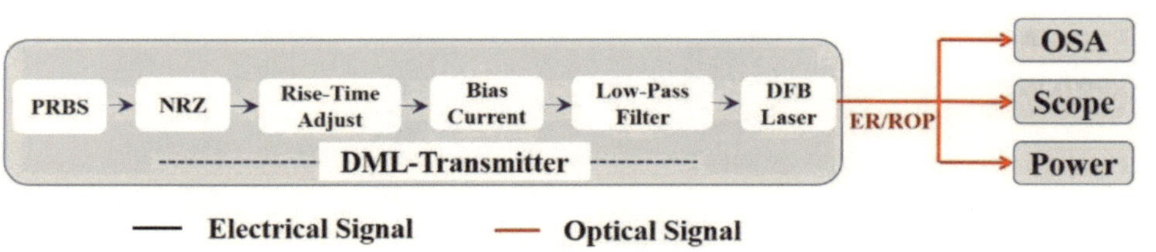

Figure 3.
Block diagram of a DML characterization bench.

Simulation Parameters	Value
BitRate	10 Gb/s
Wavelength	1.57 μm
Laser Bias Current	90 mA
Laser Confinement Factor	0.5
Henry Linewidth Enhancement Factor	3.0

Table 1.
DML simulation parameters.

A non-return to zero (NRZ) electrical modulation scheme is supplied with pseudorandom binary sequence through modified Wichman-Hill generator with length M bits = TimeWindow*Bit-Rate, which is applied on the DFB laser while biasing at given current in order to modulate its amplitude [27]. The transmitted optical pulse after modulation for the giving rate equation laser is expressed by Eq. (1).

$$A(t)_{DML} = I_{bias} + \sum_{k=-\infty}^{\infty} A_D I_{pulse}(t - kT) \tag{1}$$

Where $A(t)_{DML}$ is the modulated optical pulse from the DML transmitter, I_{bias} is the bias current, A_D is the PRBS data sequence of the form zeros and ones, each indexing with k, I_{pulse} is the encoding signaling format used (return to zero (RZ), NRZ, etc.), and T the bit period.

The received optical power (ROP) and ER at different laser modulation current and bias are depicted in **Figure 4**. In section 3, this obtained result will be optimized through our simplified HM approach.

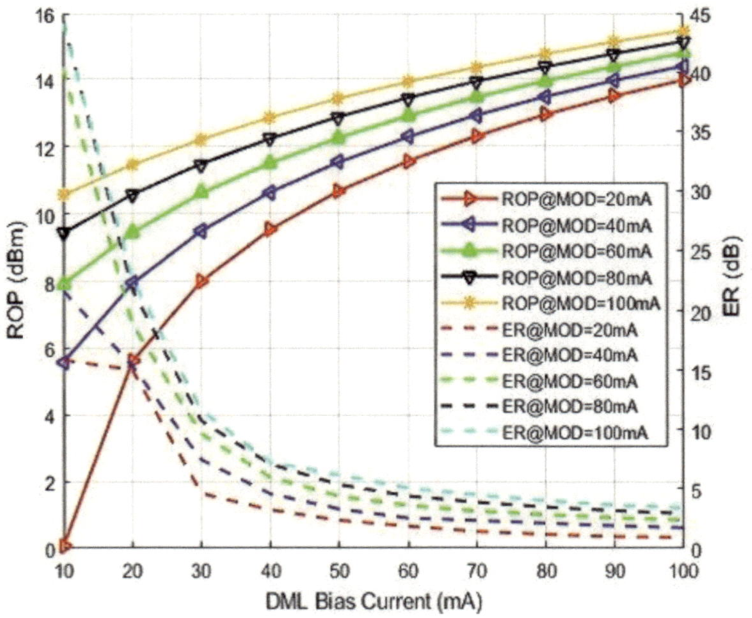

Figure 4.
ER and ROP variation against DML biasing current.

2.2 EML signal generation

EML signal generation as an alternative to DML can eliminate significantly the frequency chirp effect associated with direct modulation scheme [9]. The procedure requires a continuous wave laser (CW-laser) providing constant optical signal into the external modulator and an external electrical signal is applied to manipulate any of the desired properties (intensity, phase and frequency) of the light. A standard EML operation mode is depicted in **Figure 5**, comprising the CW laser, external modulator and the external electrical driving voltage.

As mentioned in the introduction section, two main approaches of external optical modulation are: i) the electro-refractive (LiNBO3 MZM); and ii) the electro-absorption (EAM) [13]. We shall briefly discuss design approaches and the mode of operations of these devices.

2.2.1 MZM

The overall behavior of the MZM as presented in **Figure 6** largely depends on its design and configuration, e.g., based on the lithium niobate crystal. A good design MZM has high ER with low chirp, which requires high driving voltage. The amount

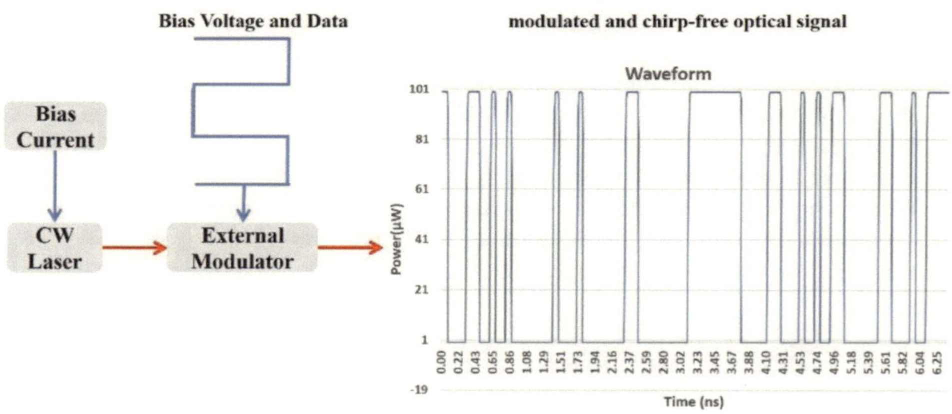

Figure 5.
Block diagram of an external modulation laser.

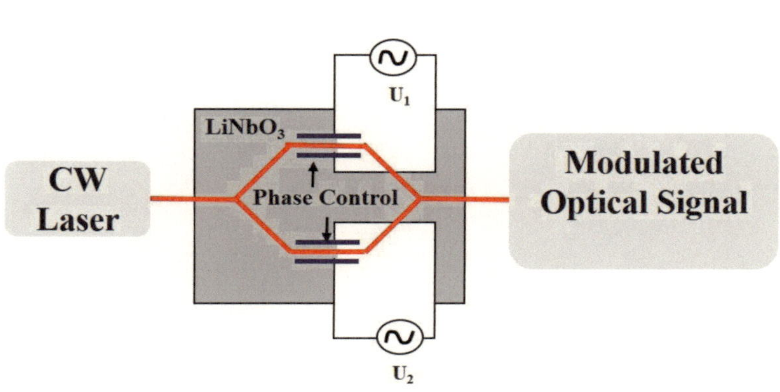

Figure 6.
Typical mode of operation of a MZM modulation scheme.

of driving voltage will then result in a large dependence of device's efficiency [22]. This voltage effect can be translated by MZM power transfer function in Eq. (2).

$$P_O(t) = \alpha_M P_i \cos\left\{\frac{\pi V(t)}{2V_\pi}\right\} \quad (2)$$

Where $P_O(t)$ is the transmitted optical power from the output of the interferometer, α_M is the modulator total IL, P_i is the input optical power from the CW laser to the modulator, $V(t)$ is the time-dependent externally applied electric voltage, and V_π the driving voltage to exert a π phase shift on the light wave carrier along the optical path of the modulator. The major drawback comparing with EAM modulation schemes is its bigger size and high driving voltage requirements, which makes it more expensive to design and with larger footprint in the integrated circuit [9].

2.2.2 EAM

EAM is considered an attractive modulation approach for fast-speed optical communications due to its low driving voltage, high bandwidth, high modulation efficiency and the possibility of monolithic integration with other semiconductor devices [28, 29]. EAM is an intensity modulator that changes the absorption properties of the carrier optical signal through the application of voltage V(t) around the band edge of the waveguides [9, 30].

Unlike MZM that modulates both intensity and phase of the carrier signal, EAM as an intensity modulator shows additional advantages, e.g., its linearity in the amplitude multilevel modulation, which offers lower total harmonic distortion when comparing to the MZM [9]. Nevertheless, MZM can achieve higher ER, an EAM with similar ER would imply increase size and therefore increase IL [9, 14].

The driving of EAM is attained with negative bias voltage to guarantee an efficient light absorption of the modulator [9]. To work in the linear region of the EAM we choose the driving voltage range [-4 V to −1.5 V], see **Figure** 7.

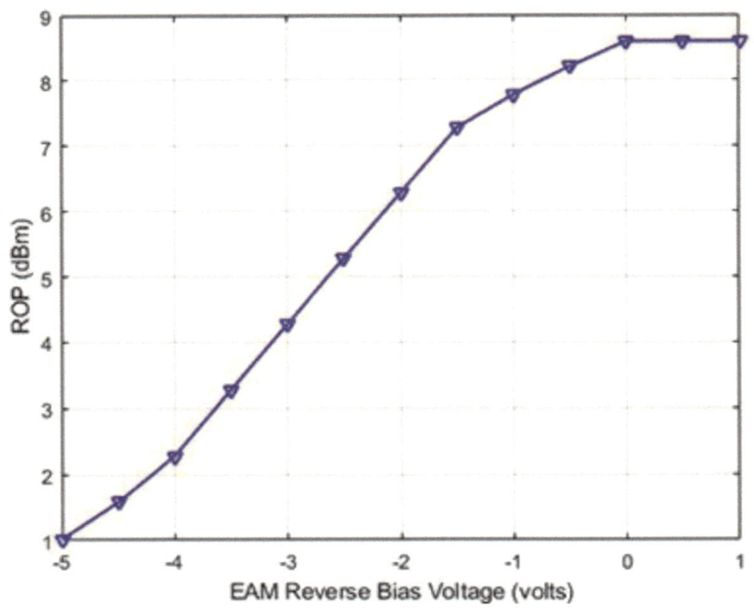

Figure 7.
EAM ROP vs. V curve. Linear region can be spotted between −4 V and − 1.5 V (laser power = 10 dBm).

Biasing the modulator must be kept within the linear region to prevent signal distortion [30]. However, voltage biasing here presents a trade-off between modulation efficiency (eye-opening) and the EAM linearity limit. With higher amplitude, EAM linear region can be extended but this will sacrifice signal efficiency with the distorted eye. In our case for 50 μm long EAM, the bias voltage is fixed at −3 V with a linear region between −4 V and − 1.5 V and the voltage swing of 1.5 V.

In the design of EAM, two major approaches are employed: i) a bulk process through Franz-Keldysh Effect, and ii) a Multi-Quantum Well (MQW) through Quantum Confined Stack Effect (QCSE) [9, 31]. Investigations show that MQW-EAM is preferred over Bulk-EAM due to its large absorption coefficient [16, 28, 31–34], leading to higher ER [35].

Considering EAM parameters from published studies [2, 32, 35–38], we simulated the modulation amplitude transfer function T(t). EAM inherent properties used as a figure of merit for a wavelength of 1.57 μm and EAM of 200 μm in length, were 0.055 dB of IL and 0.115 dB of ER per 1 μm length of EAM. Therefore, an increase of EAM ER (and thus EAM length) is given at the expense of an IL increase. This is a major setback for the use of EAM in EML modulation schemes.

Furthermore, we simulated T(t) parameters interpolation to mimic the behavior of 200 μm EAM and obtained T(t) against V(t) for different EAM length at 1.57 μm optical wavelength based on the properties of EAM length stated above.

The obtained results of EAM length versus its IL and ER are provided in **Figure 8**.

The compromise between EAM IL and ER, by changing EAM size, allow us to reach the necessary requirements of our optical transmitter, essential information to model our hybrid transmitter as presented in the section 3.

Operation of EML through the EAM intensity modulator stated here can be expressed by Eq. (3).

$$A(t)_{EML} = T(t).A(t)_{cw} \tag{3}$$

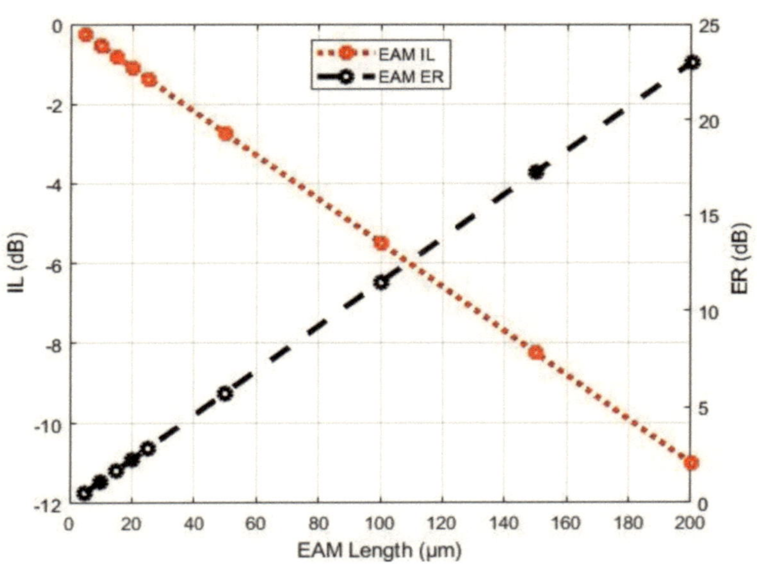

Figure 8.
EAM ER and IL versus EAM lengths. For the ER values, the EAM was biased at −4 V and modulated at 0 V.

Where $A(t)_{EML}$ is the modulated optical signal output from EAM, T(t), the complex field-intensity transmission function which depends on the length of EAM and the bias voltage while $A(t)_{cw}$ is the constant optical carrier signal supplied by the CW laser. T(t) can be specified in dB by squaring its magnitude giving by

$$T(t)_{dB} = 10.\log\left(|T(t)|^2\right) \tag{4}$$

The modulating voltage v(t) and the bias voltage V_o which are used for electrical driving of the modulator are related to T(t) according to Eqs. (5) and (6).

$$V(t) = V_o + v(t) \tag{5}$$

$$T(t) = T(t)_{dB} \cdot V(t) \tag{6}$$

3. Hybrid modulation (HM)

An obvious treat on conventional communication procedures with the current BB which leads to high system impairment and overloads has led scientific community to think further on the prospect of hybrid combination of processes and components [39]. This approach can be implemented monolithically in an InP-based platform, allowing higher spectral efficiency and the ability to generate high bit-rate signals [40].

Different limitation can be addressed for DML and EML modulation schemes. For instance, DML can only be used for relatively lower speed ≤25 Gb/s and cannot be transmitted beyond 10 km for data-center interconnect (DCI) and passive optical networks (PON) systems [7]. On the other hand, external modulation through EAM is limited by high IL values to address requirements of high ER [2]. Combining the advantages versus limitations of these two modulation schemes can result in improvements of the overall signal generation for DCI and PON systems, by mitigating the problems of transmission loss and group velocity dispersion (GVD).

Past works on optimization of transmission efficiency are associated with costly and complex system designs [22–25]. Hence, a procedure to reduce or eliminate these limiting factors are vital to the success of 5G deployment. Since PIC is still undergoing its maturity stage, its best design, functionality and efficiency are still under research. Several approaches used for impairment compensation can be optimized through hybrid combinations of BBs. In this work, we have conducted extensive studies and simulations of procedures for signal generations through direct and external modulation schemes. DML limitations have been discussed in section 2.1. External modulation on the other hand presents an attractive alternative to DML although, nevertheless with the constrain of high IL, which requires a laser with enough output power to overcome this loss.

Thus, HM concept appears as a useful approach, by exploring the advantages of both direct and external modulation approaches and components functionalities to produce a full integrated system. One major obstacle in the proposed hybrid combination is the presence of short noise dominated by photons from the DML due to spontaneous emission and the electron–hole recombination which significantly reduces the signal-to-noise ratio (SNR) of the hybrid transmitter [9]. We have provided a measure to reduce this noise and also reduce the transient chirp in the DML pulse through optical signal reshaper (OSR) approach which will be presented later in this section.

3.1 HM-model

Modulated optical signal from DML is launched into the optical signal input of the characterized EAM under study in order to re-modulate the optical signal for intensity signal generation. The T(t) earlier described in Eq. (6) that depends on the driving bias voltage and length of the modulator is loaded as a data file through VPIphotonics transmission maker optical simulator to control the EAM [41]. Therefore, for HM-model, Eq. (3) can be rewritten as presented in Eq. (7)

$$A(t)_{HM} = T(t) \cdot A(t)_{DML} \tag{7}$$

From the Eq. (3) We have replaced the $A(t)_{cw}$ with $A(t)_{DML}$ obtained from Eq. (1) which is the modulated optical signal launched from the DML and $A(t)_{HM}$ is the output of the hybrid transmitter.

$T(t)$ enables us to control and decide on the length of the modulator that is sufficient to guarantee the desire ER and with the biasing of the DML current, we can adjust the power budget to compensate the IL imposed by EAM as a result of modulator length.

The corresponding schematic is presented in **Figure 9**. It consists of PRBS and NRZ electrical signal encoder used for driving DML. In the case of HM, the electrical signal is split into two to drive both DML and EAM. the EAM amplitude is configured with reverse bias voltage of −3 V which fall within the EAM linear region.

The obtained signal $A(t)_{EAM}$ can be optimized with high ER and power budget sufficient for DCI and PON. The optimization of our hybrid modulator was done with optical wavelength at 1.57 μm where we obtained the parameters for EAM characterization.

The measured ER and ROP by changing the length of EAM and modulation current at fixed bias (90 mA) current of DML are presented in **Figure 10** and **Figure 11** respectively. The DML ROP and ER before launching the optical signal for intensity modulation are also highlighted on the graphs.

We further launched the optical signal at different EAM lengths tested into optical distribution networks (ODN) ranging from 5–40 km in order to study the behavior of the signals and responses to dispersion and non-linearity in the fiber, see **Figure 12**. With increased EAM length, the signals present an improved error rate from the 5 km starting point up to 20 km which is attributed to high ER.

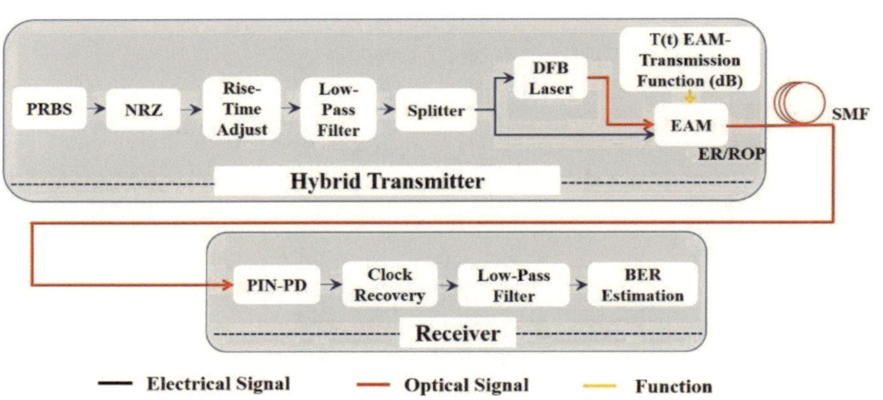

Figure 9.
Simulation setup for HM approach.

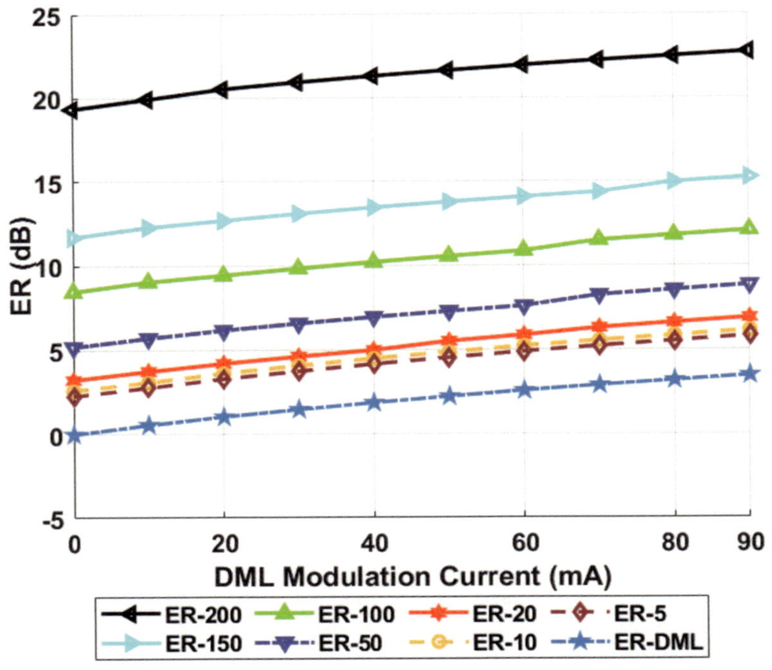

Figure 10.
ER at hybrid transmitter's output for different EAM-length against DML modulation current: EAM bias, V_o *= -3 V, EAM voltage swing,* $v(t)$ *= 2 V, DML Bias = 90 mA.*

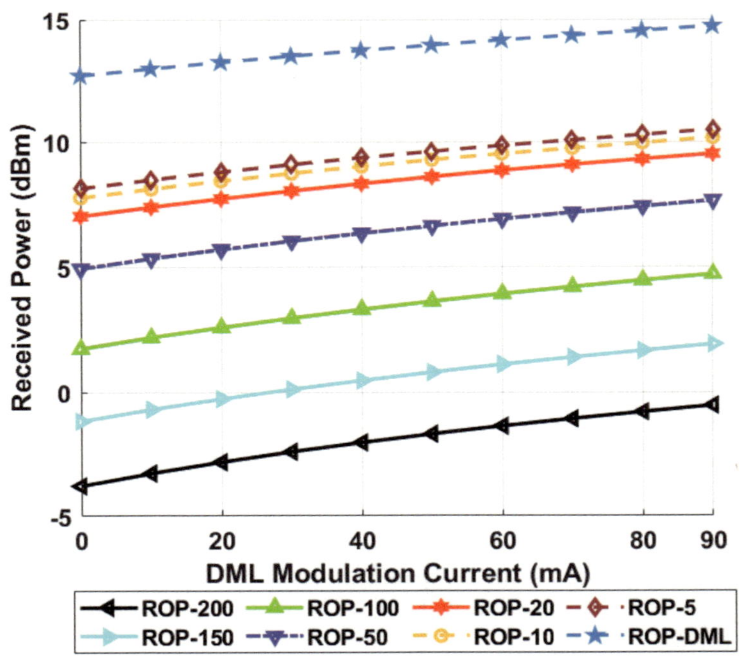

Figure 11.
ROP at hybrid transmitter's output for different EAM-length against DML modulation current: EAM bias, V_o *= -3 V, EAM voltage swing,* $v(t)$ *= 2 V, DML Bias = 90 mA.*

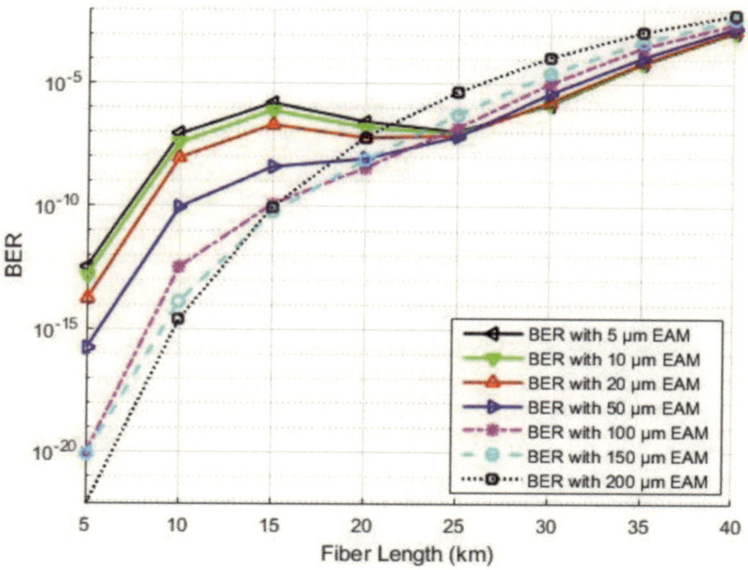

Figure 12.
BER and ROP versus fiber length for optical signals stemming from EAM with different lengths.

However, since higher ER also contributes to higher IL coupled with the attenuation in the fiber, the signal with lower EAM length (EAM = 5 μm) gives a better error rate than when EAM length is 200 μm as we continue to increase the length of the fiber as presented in **Figure 12**. At 40 km all the signals from different EAM lengths tested were received at an error rate less than 10^{-3}.

3.2 Simplified high bit-rate multilevel signal generation

Short reach fiber optic communication such as PON, short-reach video on demand (VoD) and DCI have recently advanced the demand for bandwidth [42]. This demands more efficient and advanced high spectral modulation format to replace the conventional NRZ line code [43] in order to guarantee high data rate beyond 50 Gb/s per channel [41, 44]. IEEE 802.3 group quad small form-factor pluggable (QSFP) 400 Gb/s specifications for short-reach data center communication systems have been studied progressively. Better data rate usage such as 56 Gb/s per channel can help to reduce system design complexity and cost [44–46]. Studies show that access, aggregation and core networks bandwidth demand in 5G can be adequately guaranteed with 400 Gb/s PAM-4 signal generations with either 56 Gb/s 8-channel or 100 Gb/s 4-channel networks [47, 48]. With this provision, some key requirements of 5G networks such as low cost, high performance and high bandwidth can be adequately guaranteed.

PAM-4 is a multi-order modulation approach which presents a bit-rate twice of the NRZ signal line code under the same baud-rate using four levels for signal transmission with two bits of logical information per each clock period [48]. However, PAM-4 presents a high degree of complexity in the signal generation, coupled with its sensitivity to amplitude noise which leads to a high signal to noise ratio (SNR) [45]. This is because PAM-4 signal have four levels with three eyes, which implies that, its signal is generated with an amplitude (A) of A/3 compared to NRZ that has a SNR = A. With this deficiency, PAM-4 signal is at least three times more sensitive to amplitude noise than NRZ.

To generate a PAM-4 signal for optical communication however, some complex system designs are used. In [48], physical coding sub-layer (PCS) is used to support forward error correction (FEC) at both transmitter and receiver for signal coding/encoding, scrambling/descrambling, signal alignment, signal sorting and control. Another important signal efficiency enhancement procedure is the use of electrical digital to analog converter (DAC) at the transmitter and receiver [43]. DAC usage also comes with some degree of nonlinearity and power greediness that can limit their usefulness for multi-level modulation scheme like PAM-4 in a cost effective system [48]. If nonlinearity of signal is not eliminated or reduced, it usually leads to signal distortion, which will require additional pre-distortion management procedures such as static pre-distortion (SPD) and dynamic pre-distortion (DPD) for non-linearity compensation [49]. The digital signal processing (DSP) for distortion and dispersion compensation as presented in [49, 50] to eliminate power fading high-frequency signals have further complexity in conventional PAM-4 signal generations that contradict 5G requirements for low cost signal generation approaches.

In [51], feedforward equalizer (FFE) and decision feedback equalizer (DFE) are employed at both transmitter and receiver to cancel the multi-level signal ISI and this is accomplished with system complexity and cost. To mitigate some of the complexity in PAM-4 signal generations in order to meet the demand for low cost, energy-efficient and low footprint demand for 5G networks, optical DAC PAM-4 signal generation was used in [44] with high tolerance to modulation nonlinearity. In terms of distance covered also, modulation through DML with high bit-rate signal shows several limitations due to the high chirp associated with its signal coupled with the lagging of lower PAM-4 eye while transmitting with high bit-rate and high modulation current [46]. Although, DML approach shows the simplest and most cost-effective measures but highly limited to low bit-rate and shot reach.

However, in our approach to eliminate these aforementioned limitations in PAM- 4 signal generation, unlike the conventional signal generation with EAM where electrical PAM-4 signal is employed to generate optical PAM-4 signal, the concept of HM approaches for NRZ signal generation we presented in [2] is further employed to generate optical PAM-4 signal. The approach is optimized in a simplified way to generate the signal for short reach transmissions without signaling, nonlinearity and CD compensation. The complete simulation setup is presented in **Figure 13**. We conducted further optimization of our HM model to design 28-GBaud PAM-4 signal and in fact, the same approach was tailored towards

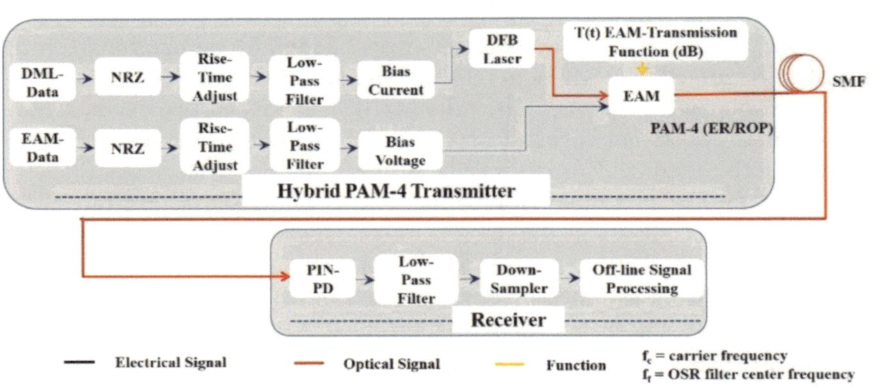

Figure 13.
Schematic of the simplified multilevel (PAM-N) signal generation approach through HM model.

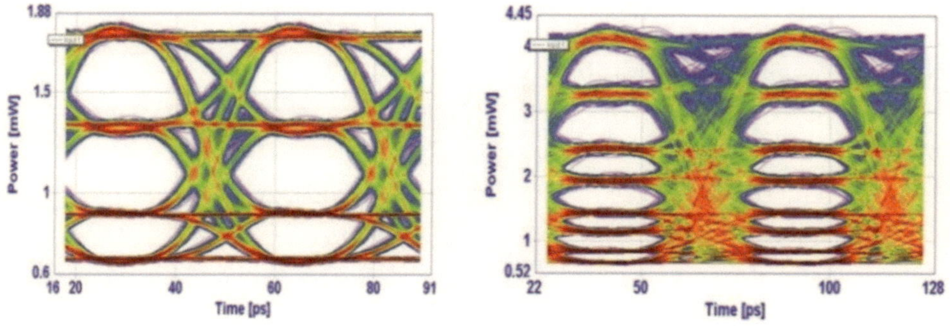

Figure 14.
Eye diagram of 28 GBaud PAM-4 and 20 GBaud PAM-8 signals generated with simplified HM approach. Here the PAM-4 signal has an ER = 4.5 dB and PAM-8 signal with ER = 7.5 dB, both signals having ROP higher than 4 dBm.

generating a simplified 20-GBaud PAM-8 signal eliminating the conventional complexity of electrical signal coding and pulse shaping.

Both transmitters show an optical launch power of more than 4 dBm with 5 dB and 7.5 dB ER respectively. At the receiver, the multilevel signals are decoded through direct detection approach with a PIN photodetector. The eye diagrams of both PAM-4 and PAM-8 signals generated with this approach are presented in **Figure 14**. The results of transmission over 8 km of both the 28 GBaud PAM-4 and 20 GBaud PAM-8 obtained through offline digital signal processing using Gaussian approximation give error analysis below 10^{-3}.

3.3 DML CML-OSR for hybrid signal compensation

The major effect of DML is the associated phase modulation to the intensity modulation, which results in CD and then ISI. In the HM model, some degree of chirp is still present in the modulated signal, which introduces limitations. With the concept of chirp managed lasers (CML) through optical signal reshaper (OSR), the transient chip from the DML can be reduced. More also, since our DML optimization earlier presented is tailored towards improving the optical signal power output and reducing the transient chirp by biasing away from the threshold. This reduces the signal ER. With the concept of CML, the ER of the DML optical signal can be optimized as well as the transient chirp. An update to the schematic in **Figure 9** is presented in **Figure 15** including a Gaussian optical filter optimized as an OSR.

The entire CML-OSR approach can be studied in [52–55]. The concept of CML decouples optical signal power and chirp of the DML signal [53], which is achieved by configuring the Gaussian filter as a band-pass filter (BPF) having a central frequency higher than the frequency of the carrier signal. This enables the filter to undergo signal edge filtering by suppressing the 1-bit while attenuating the 0-bit of the distorted NRZ optical pulse from DML. The filter is configured with a 3 dB bandwidth lower than the bandwidth of the carrier signal which distorts the signal but at the same time cut-off most of the high-frequency noises associated with the optical signal pulse. Hence, the bandwidth is highly significant to the overall performance of the OSR but care must be taken in using this approach for hybrid multilevel signal generations (e.g., PAM-4 and PAM-8) if the OSR bandwidth is further reduced or if the filter central frequency is further increased. The behavior of this approach is shown in **Figure 16** while the simulation parameter is also presented in **Table 2**. Applying this model on our generated HM optical signal

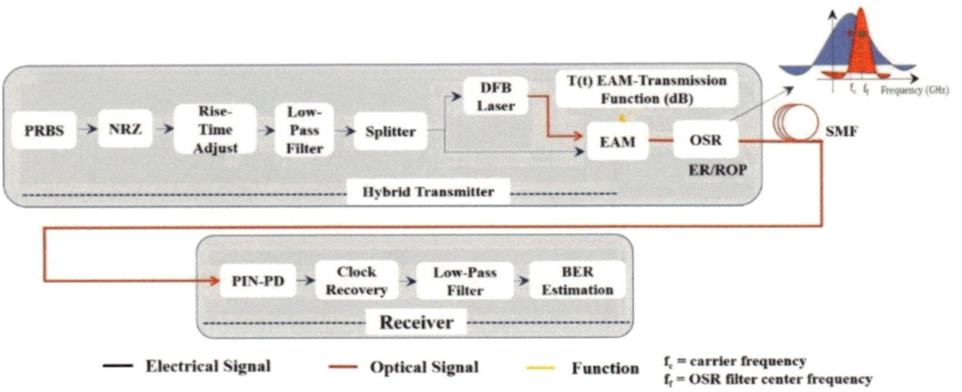

Figure 15.
Hybrid simulation approach + CML-OSR.

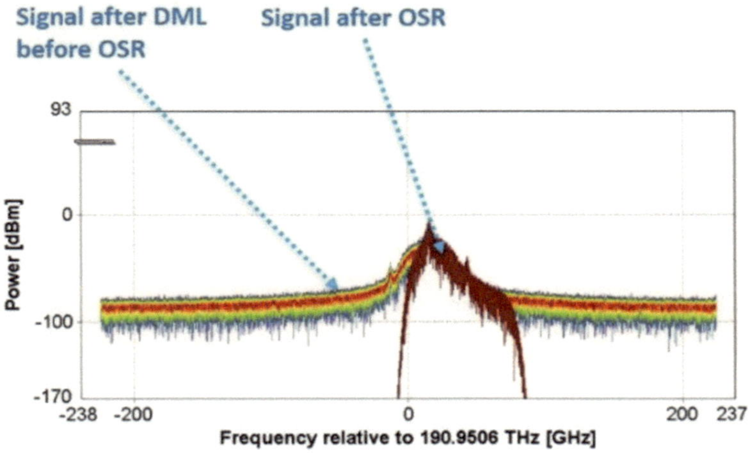

Figure 16.
CML-OSR showing DML signals before and after OSR filtering.

Filter Parameters	Value
Filter Type	BandPass
Transfer Function	Gaussian
Filter Center Frequency	190.99e12 Hz
Filter Bandwidth	60 GHz
Gaussian Order	4

Table 2.
CML-OSR simulation parameters.

improves the signal ER which further improves the overall performance of the final optical signals from the hybrid transmitter. More also, the filter significantly reduced the transient chirp associated with the DML signal. We applied this concept to both the binary and multilevel signals generated in our earlier sections and significant improvement in terms of reach was observed. With OSR, we were able to transmit a binary 40 Gb/s HM signal beyond 40 km and both 28 GBaud PAM-4 and 20 GBaud PAM-8 signals respectively up to 10 km.

4. Conclusion

We have proposed and demonstrated the concept of simplified high bit-rate signal generation with a HM approach in this chapter. The current 5G and the beyond technologies specifically target such a model with less complexity but unprecedented spectral efficiency in order to reduce the capital expenditure (CAPEX) and operation expenditure (OPEX) of signal transmission and at the same time guarantee the speed requirements of the application over optical communication networks. Through simulations, we have demonstrated and shown that there is a clear path to achieve 5G backhauling without the need for CD pre and post compensation for high bitrate signal generation in short and medium reach networks. We also showed that proper optimization can improve signal launch power and eliminate the necessity of expensive optical power amplifier for high bitrate signal transmission. HM clearly simplified signal generations as we have presented in this chapter with right combination of process and components. Further research is ongoing to implement this model on a PIC so that we can perform a real-life laboratory test of the chip and investigate other areas of optimizations.

Acknowledgements

This work is supported by the project Virtual Fiber Box, with reference number POCI-01-0247- FEDER-033910, funded by the European Regional Development Fund (FEDER), through the Operational Program Competitiveness and Internationalization (COMPETE 2020), of Portugal 2020 framework (P2020).

Author details

Adebayo E. Abejide[1,2*], Madhava R. Kota[1,2], Sushma Pandey[1,2], Oluyomi Aboderin[1], Cátia Pinho[1,3], Mário Lima[1,2] and António Teixeira[1,2]

1 Instituto de Telecomunicações (IT), University of Aveiro, Aveiro, Portugal

2 Department of Electronics, Telecommunications and Informatics (DETI), University of Aveiro, Aveiro, Portugal

3 PICadvanced, PCI-Creative Science Park via do Conhecimento, Edifício Central, Ílhavo, Portugal

*Address all correspondence to: adebayo@ua.pt

References

[1] Chowdhurry, Mostafa Zaman, et al. "The Role of Optical Wireless Communication Technologies in 5G/6G and IoT Solutions: Prospects, Directions and Challenges." Applied Sciences 9.20 (2019): 4367.

[2] Adebayo E. Abejide, et al. "Hybrid Modulation Approach for Next Generations Optical Access Networks." Proceedings of CSNDSP Portugal, July 2020.

[3] E. Saini, R. Bhatia and S. Prakash, "High speed broadband communication system for moving trains using Free Space Optics," *2016 International Conference on Computational Techniques in Information and Communication Technologies (ICCTICT)*, New Delhi, 2016, pp. 47–50.

[4] A Teixeira, "Photonic Integrated Circuits for NG-PON2 ONU Transceivers" Optical Fiber Communication Conference (OFC), 2019.

[5] Christophe Peucheret, "Direct and External Modulation of Light". DTU Fotonik, Department of Photonic Engineering, Technical University of Denmark, Denmark. November, 2009.

[6] Fumio Koyama, Kenichi Iga, "Frequency Chirping in External Modulators". Journal of Lightwave Technology, Vol. 6, No. 1, January 1988.

[7] EML vs DML for Datacenter and Client Side Transceivers. [Online]. Available: https://www.neophotonics.com.

[8] Robbert Van der Linden. "Adaptive Modulation Techniques for Passive Optical Networks". Doctorate Thesis, Eindhoen University of Technology, Netherlands. April, 2018.

[9] Le Nguyen Binh. "Optical Fiber Communication with Matlab and Simulink Models. Second Edition". Taylor and Francis Group, LLC, 2015.

[10] Ning Hua Zhu Et al. "Direct Modulated Semiconductor Lasers". IEEE Journal of Selected Topics in Quantum Electronics. Vol. 24, No. 1, January 2018.

[11] Z. F. Fan, "Chirp Managed Lasers: A New Technology for 10Gb/s Optical Transmitters," pp. 39–41, 2007.

[12] International Telecommunication Union ITU-T. "PON transmission technologies above 10 Gb/s: Transmission Systems and Media, Digital Systems and Networks". Series G, Supplement 64, February, 2018.

[13] H. Venghaus and N. Grote, "Fibre Opti Communication: Key Devices". Second Edition. Springer International Publishing, Switzerland, 2017.

[14] Govind P. Agrawal, Fiber-Optics Communication Systems, Third Edition.

[15] Dimension- "Directly Modulated Lasers on Silicon: Deliverable Report D 2.1", Report on System Specifications, Research and Innovation Actions (RIA), H2020-ICT-27-2015 Photonics KET. July, 2016.

[16] Juan Camilo Velasquez Micolta, "Next Generation Optical Access Networks and Coexistence with Legacy PONs", Universitat Politecnica De Catalunya, Barcelona. July, 2019.

[17] Hidekazu et al, "EAM-Integrated DFB Laser Modules with more than 40-GHz Bandwidth", IEEE Photonics Technology Letters. Vol.13, No. 9, September, 2001.

[18] Yu, J., Jia, Z., Huang, M. F., Haris, M., Ji, P. N., Wang, T., & Chang, G. K. (2009). Applications of 40-Gb/s chirp-managed laser in access and metro networks. Journal of Lightwave Technology, 27(3), 253–265.

[19] Tomkos, Ioannis & Chowdhury, D. & Conradi, J. & Culverhouse, D. & Ennser, K. & Giroux, C. & Hallock, B. & Kennedy, T. & Kruse, A. & Lascar, N. & Roudas, I. & Sharma, M. & Vodhanel, R.S. & Wang, C.-C. (2001). Demonstration of negative dispersion fibers for DWDM metropolitan area networks. Selected Topics in Quantum Electronics, IEEE Journal of. 7. 439–460. 10.1109/2944.962268.

[20] i, Philip N., Jianjun Yu, Ting Wang, Xueyan Zheng, Yasuhiro Matsui, Daniel Mahgerefteh, Kevin McCallion, Zhencan Frank Fan, and Parviz Tayebati. "Chirp-managed 42.8 Gbit/s transmission over 20 km standard SMF without DCF using directly modulated laser." In 33rd European Conference and Exhibition of Optical Communication, pp. 1–2. VDE, 2007.

[21] dokoro, Takashi & Kobayashi, Wataru & Fujisawa, Takeshi & Yamanaka, Takayuki & Kano, Fatima. (2011). High-Speed Modulation Lasers for 100GbE Applications. Optics InfoBase Conference Papers. 10.1364/OFC.2011.OWD1.

[22] Troppenz, U. & Kreissl, Jochen & Moehrle, Martin & Bornholdt, C. & Rehbein, W. & Sartorius, B. & Woods, Ian & Schell, Martin. (2011). 40 Gbit/s Directly Modulated Lasers: Physics and Application. Proceedings of SPIE - The International Society for Optical Engineering. 7953. 10.1117/12.876137.

[23] M. Shahin, K. Ma, A. Abbasi, G. Roelkens, G. Morthier, 45 Gb/s Direct Modulation of Two-Section InP-on-Si DFB Laser Diodes, IEEE Photonics Technology Letters, 30(8), p.685–687 doi:10.1109/LPT.2018.2811906 (2018).

[24] K. Otsubo, M. Matsuda, S. Okumura, A. Uetake, M. Ekawa and T. Yamamoto, "Low-driving-current high-speed direct modulation up to 40 Gb/s using 1.3-μm semi-insulating buried-heterostructure AlGaInAs-MQW distributed reflector (DR) lasers," 2009 Conference on Optical Fiber Communication - incudes post deadline papers, San Diego, CA, 2009, pp. 1–3.

[25] M. C. Wu, C. Chang-Hasnain, E. K. Lau and X. Zhao, "High-Speed Modulation of Optical Injection-Locked Semiconductor Lasers," OFC/NFOEC 2008–2008 Conference on Optical Fiber Communication/National Fiber Optic Engineers Conference, San Diego, CA, 2008, pp. 1–3, doi: 10.1109/OFC.2008.4528483.

[26] Chan, Chun-Kit, Wei Jia, and Zhixin Liu. "Advanced modulation format generation using high-speed directly modulated lasers for optical metro/access systems." In 2011 Asia Communications and Photonics Conference and Exhibition (ACP), pp. 1–12. IEEE, 2011.

[27] R Coates, G. Janacek, and G. Lever, "Monte Carlo Simulation and Random Number Generation", IEEE J. Selected Areas Communication, Vol. 6, Pp. 58–66, 1988.

[28] Rashed, Ahmed Nabih Zaki. "Recent developments and signal processing of low driving voltage and high modulation efficiency electro-absorption modulators (EAMs)." International Journal of Image, Graphics and Signal Processing 4, no. 4 (2012): 11.

[29] M Theurer et al, "56Gb/s L-band InGaAIAs Ridge Waveguide Electro-Absorption Modulated Laser with Integrated SOA", Wiley Online Library. December 3, 2015.

[30] W.-J. Huang, C.-C. Wei, and J. Chen, "Optical dac for generation of pam-4 using parallel electro-absorption modulators," in ECOC 2016; 42nd European Conference on Optical Communication. VDE, 2016, pp. 1–3.

[31] Dazeng Feng et al, "High-Speed GeSi Electro-Absorption Modulator on

SOI Waveguide Platform" IEEE Journal of Selected Topics in Quantum Electronics. November, 2013.

[32] Moritz Friedrich, "Polarization Multiplexed Photonic Integrated Circuits for 100Gb/s and Beyond" Universitat Berlin, July. 2018.

[33] Sheikhi, M. H., S. A. Emamghoreishi, S. Javadpoor, and M. K. Moravvej-Farshi. "A new theoretical design optimization of multiple quantum-well electroabsorption modulator." In IEEE/LEOS 3rd International Conference on Numerical Simulation of Semiconductor Optoelectronic Devices, 2003. Proceedings, pp. 17–18. IEEE, 2003.

[34] Smit, Meint, Xaveer Leijtens, Huub Ambrosius, Erwin Bente, Jos Van der Tol, Barry Smalbrugge, Tjibbe De Vries et al. "An introduction to InP-based generic integration technology." Semiconductor Science and Technology 29, no. 8 (2014): 083001.

[35] Mastronardi L, Banakar M, Khokhar AZ, Hattasan N, Rutirawut T, Bucio TD, Grabska KM, Littlejohns C, Bazin A, Mashanovich G, Gardes FY. High-speed Si/GeSi hetero-structure Electro Absorption Modulator. Opt Express. 2018 Mar 19;26(6):6663–6673. doi: 10.1364/OE.26.006663. PMID: 29609353.

[36] M. Baier et al., "112 Gb/s PDM-PAM4 Generation and 80 km Transmission Using a Novel Monolithically Integrated Dual-Polarization Electro-Absorption Modulator InP PIC," 2017 European.

[37] D. Feng et al., "High-Speed GeSi Electroabsorption Modulator on the SOI Waveguide Platform," in IEEE Journal of Selected Topics in Quantum Electronics, vol. 19, no. 6, pp. 64–73, Nov.-Dec. 2013, Art no. 3401710, doi: 10.1109/JSTQE.2013.2278881.

[38] M. Trajkovic, F. Blache, H. Debregeas, K. A. Williams and X. J. M. Leijtens, "Increasing the Speed of an InP-Based Integration Platform by Introducing High Speed Electroabsorption Modulators," in IEEE Journal of Selected Topics in Quantum Electronics, vol. 25, no. 5, pp. 1–8, Sept.-Oct. 2019, Art no. 3400208, doi: 10.1109/JSTQE.2019.2913727.

[39] Ding, Liang, Radhakrishnan L. Nagarajan, and Roberto Coccioli. "Compact optical transceiver by hybrid multichip integration." U.S.Patent 9,921,379, issued March 20, 2018.

[40] Zhang, Jing, Grigorij Muliuk, Joan Juvert, Sulakshna Kumari, Jeroen Goyvaerts, Bahawal Haq, Camiel Op de Beeck et al. "III-V-on-Si photonic integrated circuits realized using micro-transfer-printing." APL photonics 4, no. 11 (2019): 110803.

[41] VPIPhotonics Software, "ModulatorEA_Measured", Version 10.0, Berlin Germany, 2019 [Online]. Available: https://www.vpiphotonics.com/index.php.

[42] T. Salgals, I. Kurbatska, S. Spolitis, V. Bobrovs, and G. Ivanovs, "Research of m-pam and duobinary modulation formats for use in highspeed wdm-pon systems," in Telecommunication Systems-Principles and Applications of Wireless-Optical Technologies. IntechOpen, 2019.

[43] A. Samani, E. El-Fiky, M. Morsy-Osman, R. Li, D. Patel, T. Hoang, M. Jacques, M. Chagnon, N. Abadía, and D. V. Plant, "Silicon photonic mach–zehnder modulator architectures for on chip pam-4 signal generation," *Journal of Lightwave Technology*, vol. 37, no. 13, pp. 2989–2999, 2019.

[44] W.-J. Huang, C.-C. Wei, and J. Chen, "Optical dac for generation of pam-4 using parallel electro-absorption modulators," in ECOC 2016; 42nd

European Conference on Optical Communication. VDE, 2016, pp. 1–3.

[45] Tekronix. *"Pam-4 signaling in high speed serial technology: Test, analysis, and debug," Tektronix, 2018.*

[46] *Y. Matsui, T. Pham, T. Sudo, G. Carey, B. Young, J. Xu, C. Cole, and C. Roxlo, "28-gbaud pam-4 and 56-gb/s nrz performance comparison using 1310-nm al-bh dfb laser," Journal of Lightwave Technology, vol. 34, no. 11, pp. 2677–2683, 2016.*

[47] *K. Nakahara, R. Hirai, T. Kitatani, N. Kikuchi, T. Fukui, K. Okamoto, Y. Sakuma, K. R. Tamura, and S. Tanaka, "Superior ber transmission of 106-gb/s/lane skewless pam-4 over 10 km by utilizing 1.3-μm directly modulated ingaalasmqw bh lasers and incoherent multiplexing of two nrz signals," in* 2018 Optical Fiber Communications Conference and Exposition (OFC). IEEE, 2018, pp. 1–3.

[48] C. Mobile, C. Telecom, C. Unicom, Huawei, FiberHome, AccelinK,

[49] *Hisense, S. E. ans Source Photonics, iNNOLIGHT, K. TECHNOLOGIES, SPIRENT, MACOM, and Inphi, "50G PAM-4 Technical White Paper," Huawei, Tech. Rep., 2010.*

[50] H. Li, G. Balamurugan, M. Sakib, J. Sun, J. Driscoll, R. Kumar, H. Jayatilleka, H. Rong, J. Jaussi, and B. Casper, "A 112 gb/s pam-4 transmitter with silicon photonics microring modulator and cmos driver," in 2019 Optical Fiber Communications Conference and Exhibition (OFC). IEEE, 2019, pp. 1–3.

[51] Q. Guo, B. Hua, C. Ju, Z. Zhang, Y. Chen, Z. Tu, and X. Huang, "Experiment demonstration of im-dd based 50-gb/s pam-4 tdm-pon downstream scheme enabled by transmitter pre-emphasis and mlse," in *2018 23rd Opto-Electronics and Communications Conference (OECC)*. IEEE, 2018, pp. 1–2.

[52] A. Roshan-Zamir, T. Iwai, Y.-H. Fan, A. Kumar, H.-W. Yang, L. Sledjeski, J. Hamilton, S. Chandramouli, A. Aude, and S. Palermo, "A 56-gb/s pam-4 receiver with low-overhead techniques for threshold and edge-based dfe fir-and iir-tap adaptation in 6y5-nm cmos," IEEE Journal of Solid-State Circuits, vol. 54, no. 3, pp. 672–684, 2018.

[53] Pan, Yue, and Yanping Xi. "Monolithically integrated chirp-managed laser (CML) based on a resonant tunneling optical spectrum reshaper filter." In 2016 Asia Communications and Photonics Conference (ACP), pp. 1–3. IEEE, 2016.

[54] Zhu, Zenyuan, and Yanping Xi. "Optimized Design of Chirp Managed Lasers with Dispersion Pre-compensation by Integral Layer-Peeling Algorithm." In Asia Communications and Photonics Conference, pp. AF2A-54. Optical Society of America, 2016.

[55] Karar, Abdullah S., John C. Cartledge, and Kim Roberts. "Transmission over 608 km of standard single-mode fiber using a 10.709-Gb/s chirp managed laser and electronic dispersion precompensation." IEEE Photonics Technology Letters 24, no. 9 (2012): 760–762.

Chapter 2

MAS: Maximum Energy-Aware Sense Amplifier Link for Asynchronous Network on Chip

Erulappan Sakthivel and Rengaraj Madavan

Abstract

A real-time multiprocessor chip model is also called a Network-on-Chip (NoC), and deals a promising architecture for future systems-on-chips. Even though a lot of Double Tail Sense Amplifiers are used in architectural approach, the existing DTSA with transceiver exhibits a difficulty of consuming more energy than its gouged design during various traffic condition. Novel Low Power pulse Triggered Flip Flop with DTSA is designed in this research to eliminate the difficulty. The Traffic Aware Sense amplifier MAS consists of Sense amplifiers (SA's), Traffic Generator, and Estimator. Among various SA'S suitable (DTSA and NLPTF -DTSA) SA are selected and information transferred to the receiver. The performance of both DTSA with Transceiver and NLPTF-DTSA with transceiver compared under various traffic conditions. The proposed design (NLPTF-DTSA) is observed on TSMC 90 nm technology, showing 5.92 Gb/s data rate and 0.51 W total link power.

Keywords: network-on-chip (NoC), double tail sense amplifier (DTSA), low power pulse triggered flip flop (LPTF)

1. Introduction

NoC may be a flourishing area for designing current application like Image processing, Signal Processing multimedia, Medical applications telecommunication, and real-time task [1]. Conservative investigation mainly focuses on low power, ultra-speed, and scalability in NoC [2]. Algorithmic [3] and architectural models [4] are made and instigated into the NoC to provide additional performance improvement than current NoC design. Existing NoC designer's shows much progress on this architectural level model by introducing outside or inside sense amplifier (SA) in on-chip communication [5]. In addition to the transmitter section (TXS) with pre-emphasis capacitance (PEC) for high speed and energy reduction in on-chip communication, it requires DC bias circuits at the receiver section (RXS). To overcome this issue, voltage sense amplifier is presented and tested in 90 nm Complementary metal-oxide–semiconductor (CMOS) cross coupled module [6]. In small circuit application user can't identify the worth of voltage SA so it is refined into Double Tail Sense Amplifiers (DTSA). This DTSA with transceiver consists of PEC at the transmitter and DTSA at RXS [7]. A low power consumption model is developed and implemented in many real-time applications. Clock Gating (CG) low power design approach at RTL TSMC 45 nm CMOS application is tested in [8].

CMOS Very-large-scale integration (VLSI) design has taken us to real working chips that rely on controlled charge recovery to operate at suggestively lower power dissipation levels than their existing counterparts The Novel Low Power pulse Triggered Flip Flop with DTSA (NLPTF) is designed by using two N-type metal-oxide-semiconductor (NMOS) transistor with an inverted clock signal as an input [9]. The output is taken from the transistor and given to the P-type metal-oxide-semiconductor (PMOS) transistor. The gate output of the PMOS transistor is given to the DTSA as an input and observes the output changes and power usage [10]. In [11] the performance improvement achieved in networks with respect to Network traffic modeling based on synthetic traffic The real-time traffic data are generated and estimated in on-chip communication, according to that TE new approach introduced for Quality of service (QoS) in [12]. This proposed design we followed above Traffic generator and TE in Traffic model and NLPTF-DTSA. The reconfigurable topology is applied in on-chip networks for performance improvement in [13]. To achieve performance improvement than [7], Maximum Energy-Aware sense amplifier (MAS) circuitry is introduced which consists of Traffic Generator (TG), Traffic Estimator (TE), capturing energy recovery [14] and NLPTF-DTSA. Clock gating (CG) Concept discussed in Sakthivel et al. [15].

The rest of this Chapter is ordered as follows. Subdivision 2 addresses the NoC system model. Proposed work and its module details are discussed in subdivision 3. The proposed results of various architectures are presented in subdivision 4. Finally, the conclusion is presented in subdivision 5.

2. System model

For improved data communication in NoC, conventional transceiver consists of PEC in TXS and DTSA circuit in TXS. Schinkel et al. Transceiver for NoC's and proposed Design is shown in **Figure 1**. The use of capacitance in TXS is to shrink in power dissipation. In NoC Circuitry communication disturbance occurs because of noise and crosstalk [16]. The transceiver with DIT (differential interconnect twist) affords a high-performance perfection. Early-stage, bidirectional interconnects are used. The EM field solver is used to investigate interconnects. The CMOS with 1.2 V, 6 m technology is used for interconnects as in [7]. **Table 1** shows Conventional Strategies.

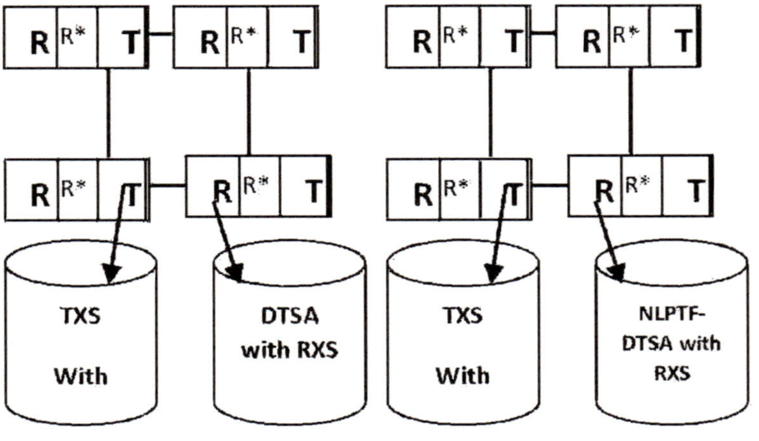

Figure 1.
Conventional and proposed Transceiver configuration.

Circuitry	EXISTING	TAS design
[7]	DTSA	TG [12], TE [12], DTSA
[18]	NLPTF	NLPTF-DTSA

Table 1.
Conventional Strategies.

3. Proposed work

Figure 2 shows a basic diagram of proposed design, which consist of following modules organized such as PEC with TXS, TE, Controller, NLPTF-DTSA, and RXS. The proposed work consists of four stages namely selection, analysis, and design and performance comparison. In the first stage of our work is suitable SA'(DTSA and NLPTF-DTSA) is selected among various sense amplifier circuitry [18] and second stage selected SA's with the transceiver (DTSA and NLPTF-DTSA) high traffic (HT) and low traffic (LT) conditions examined. In the third stage, we compared NLPTF-DTSA for complete transceiver with [7].

3.1 NLPTF-DTSA circuit

The NLPTF-DTSA [18] is as shown in **Figure 3**. NLPTF -DTSA is used to solve the troubles associated with conventional Pulsed Flip Flop (P-FF) designs. The basic procedure of NLPTF -DTSA is plummeting the number of NMOS transistors in the discharging path. The next step NLPTF -DTSA is supporting a system to enhance the strength pull down by allocation value in to "1." The new transistor stacking circuitry is opposed to transistor S2 which is distant from the discharging path. Transistor S2, in conjunction through an extra transistor N3, forms a pass transistor logic (PTL) size of two AND gates of transistor S1. Since the inputs of the two AND logic gates are matching, the output node is reserved at zero time. When input signal1 and input signal 2 are equal to "0", temporary floating at the node is basically risk-free. By the rising edge limits of the clock unit, both transistors S1 and S2 be turned resting on. This design is subsequently turned on transistor S3 by an instance width. The switching power is less at each node due to weakening voltage swing [10]. The functional diagram of NLPTF-DTSA and simulation results are shown in **Figure 3**.

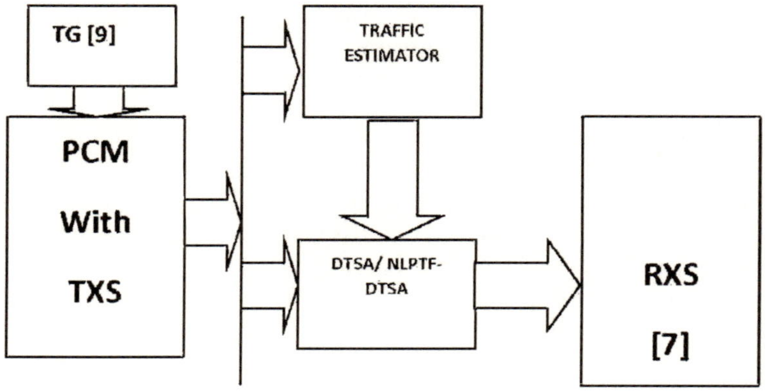

Figure 2.
Proposed Transceiver Block diagram.

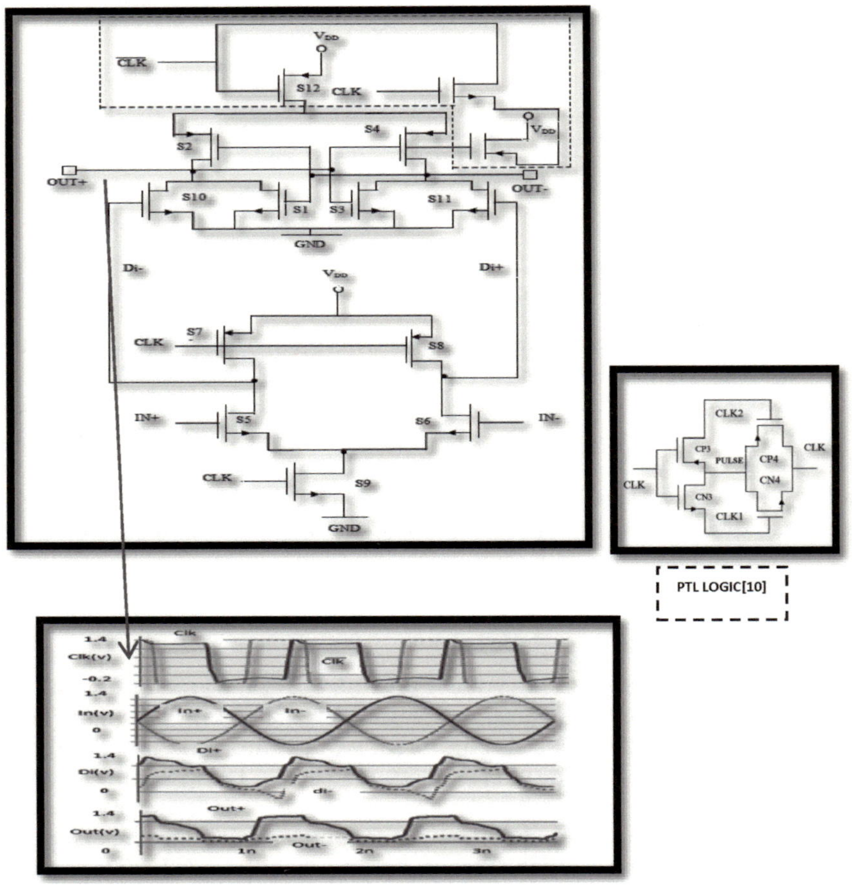

Figure 3.
(a) The functional diagram of the V-DTSA module and (b) simulation result

3.2 TXS with PEC

The technical concepts of TXS with PEC are similar to that of Schinkel et al.

3.3 Low swing transmitter

The series capacitance in transmitter is used to drive the bus and reduces the swing factor. The technical parameters of the full swing (FS), multi VDD mode (MVM) capacitive low swing transmitter (CLS) are tabulated in the **Table 2**.

3.4 Optimal swing receiver

In a transceiver circuit, SA is the best data receiver when compared to the conventional comparator [7]. To avoid transistor stack, the SA circuit is split into two tails and fed with separate supply voltage.

To gain maximum power reduction in NoC architecture, NLPTF [10] technique is implemented in DTSA module.

3.5 TG and TE

The Statistical Traffic model [12] is implemented in this approach. By which various traffic condition (image, Data) applied into SA's with Transceiver.

Modes	Inter connect	Technology	Supply voltage	Voltage swing	Driver size
FS	Shielded single ended	1.2 v, 6 metal, 90 nm CMOS 2 mm, Rwire = 400 Ω	1.2 V	1.2 V	Wn = 8 μm Wp = 20 μm
MVM	DIT		V_{DDH}= 1. 2 V V_{DDL}= 1. 08 V	120 mv	Wn = 8 μm Wp = 20 μm
CLS	DIT		1.2 V	120 mv	Wn = 1.6 μm Wp = 4 μm

Table 2.
Different modes comparison.

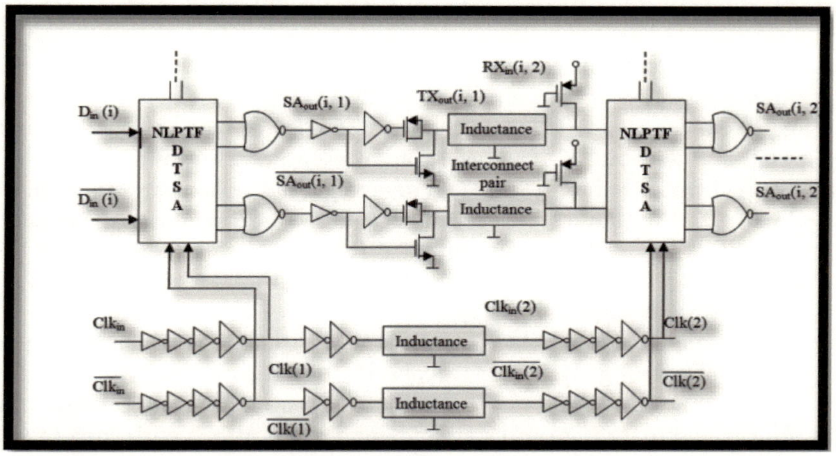

Figure 4.
The functional diagram of the Complete Transceiver Circuit.

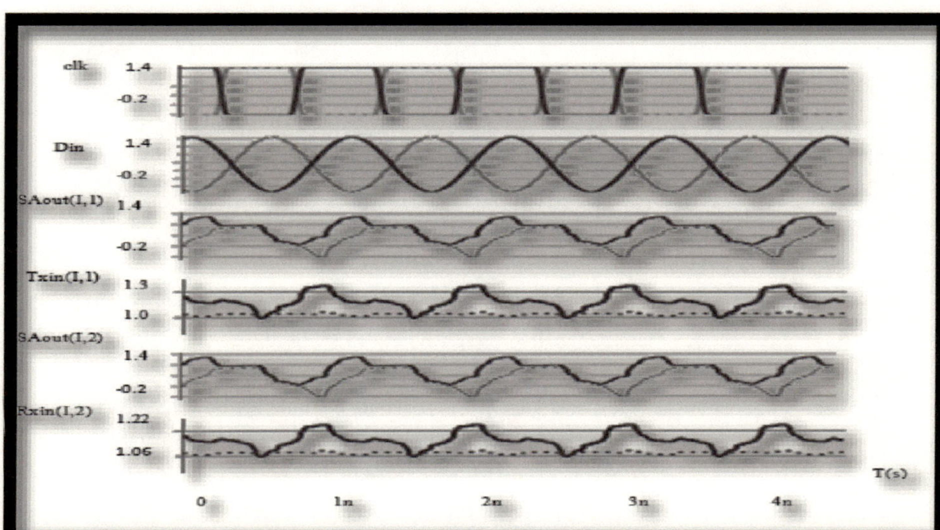

Figure 5.
Complete Transceiver simulation result.

3.6 Complete transceiver

The complete transceiver circuit is made of transmitter connected to the receiver via DIT [7]. The complete transceiver architecture is shown in the **Figure 4**. And the experimental results of complete transceiver in **Figure 5**.

4. Results and discussions

The performance parameters of the DTSA and NLPTF-DTSA with transceiver are examined using 90 nm technologies. Synopsys Design Compiler is used for Gate level net list creation. Synopsys™ Prime Power is used for Power analysis [17]. The switching factors are reported by the proposed work and examined in Intel® 3.1 GHz LGA 1155 Core i3–2100 Processor, and a system with Window Xp. The technical level similar to [7] carried out various modes such as FS, MVM, and CLS.

The NoC model synthesized code is made to evaluate 90 nm TSMC CMOS technology under the operating frequency of 500 MHZ, 1.2 V supply voltages and 0.5 switching factor. The Sleep mode and Active mode power consumption are tested.

* With CG and without CG The results are presented in **Figures 6** and **7**. It is inferred that the proposed NLPTF-DTSA gives a greater result in terms of power as related with DTSA modules such as single-ended conditional capturing energy recovery (SCCER) [10], DCCER [10], static differential energy recovery (SDER) [10], pulsed flip flop (PFF) [18], NLPTF-DTSA. A mathematical expression for technical evaluation is similar to [19].

The energy consumption, delay, data Rate and static power consumption results are presented in **Figures 8–11**. The DTSA, NLPTF-DTSA circuitry results are estimated under HT and LT. The overall comparison of various parameters (Energy consumption, delay, data Rate and static power consumption) with current work is shown in **Table 3**. The overall results of proposed design give greater results than conservative design.

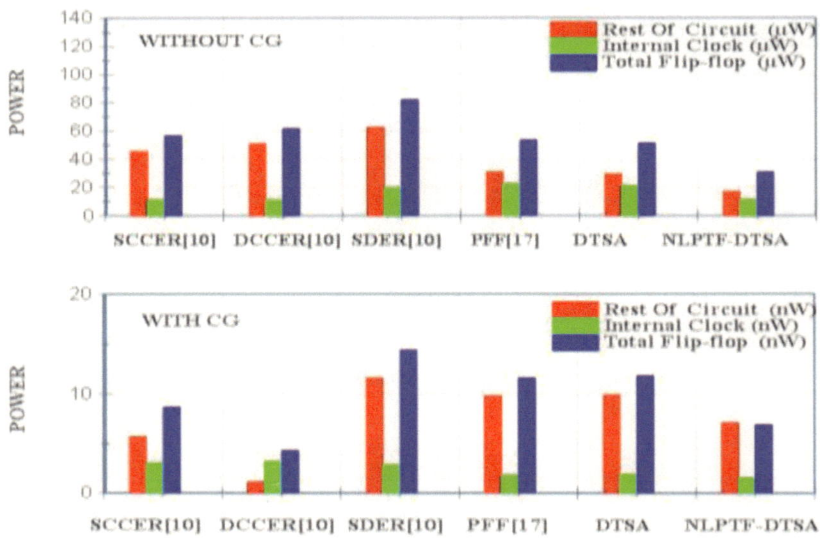

Figure 6.
Power comparison in Sleep mode.

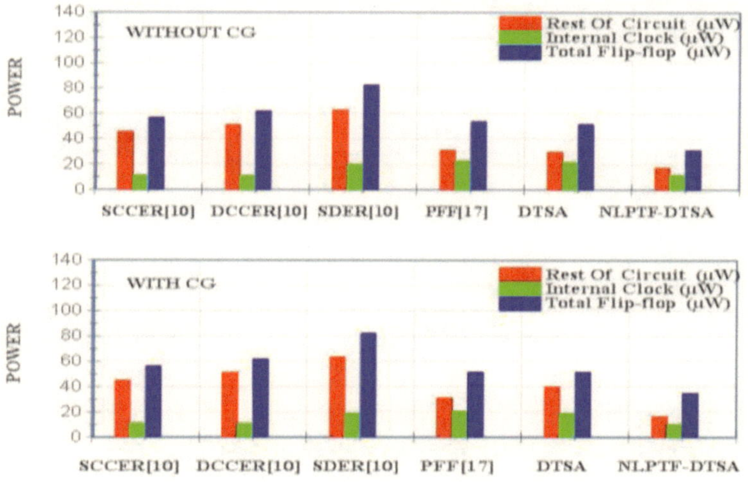

Figure 7.
Power comparison in Active mode.

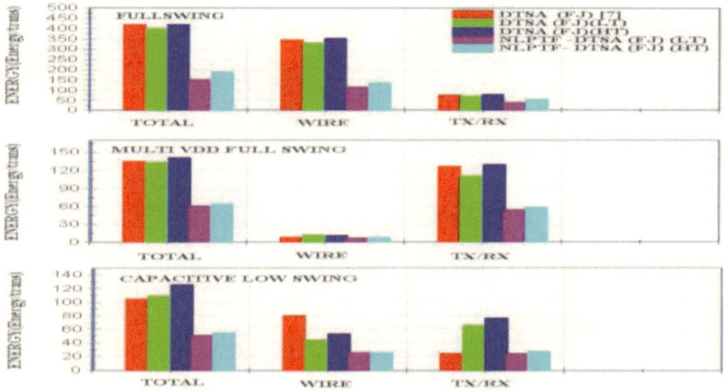

Figure 8.
Energy comparison of DTSA modules.

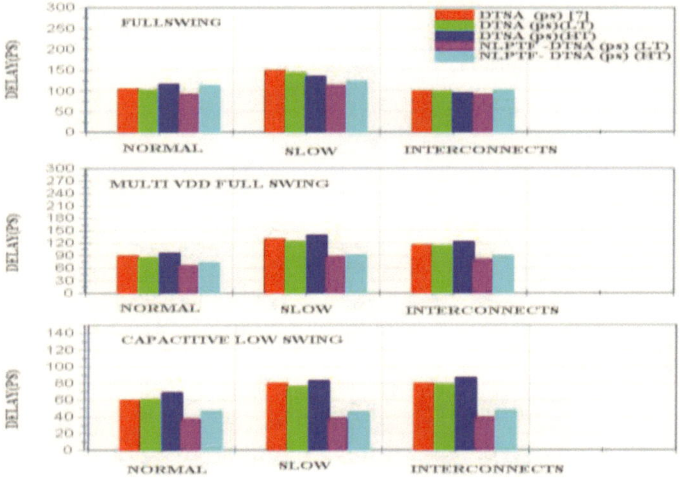

Figure 9.
Delay comparison of DTSA modules.

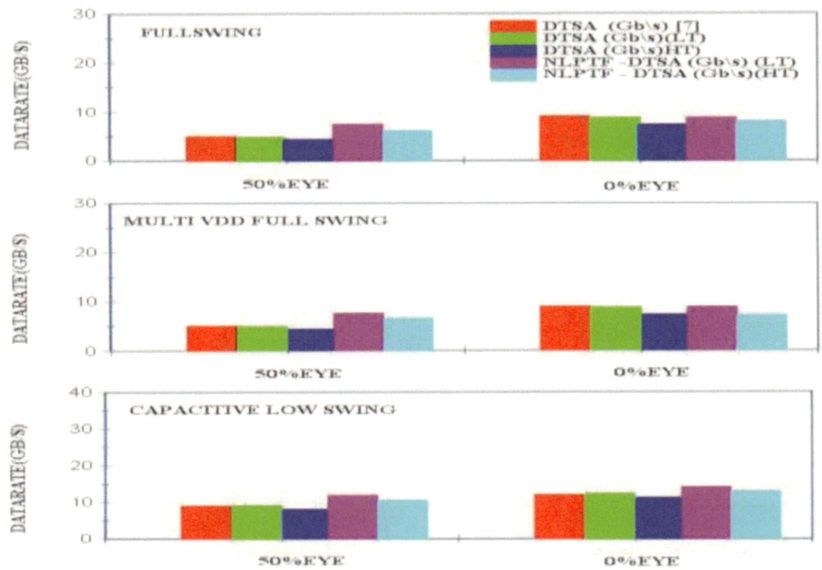

Figure 10.
Data Rate comparison of DTSA modules.

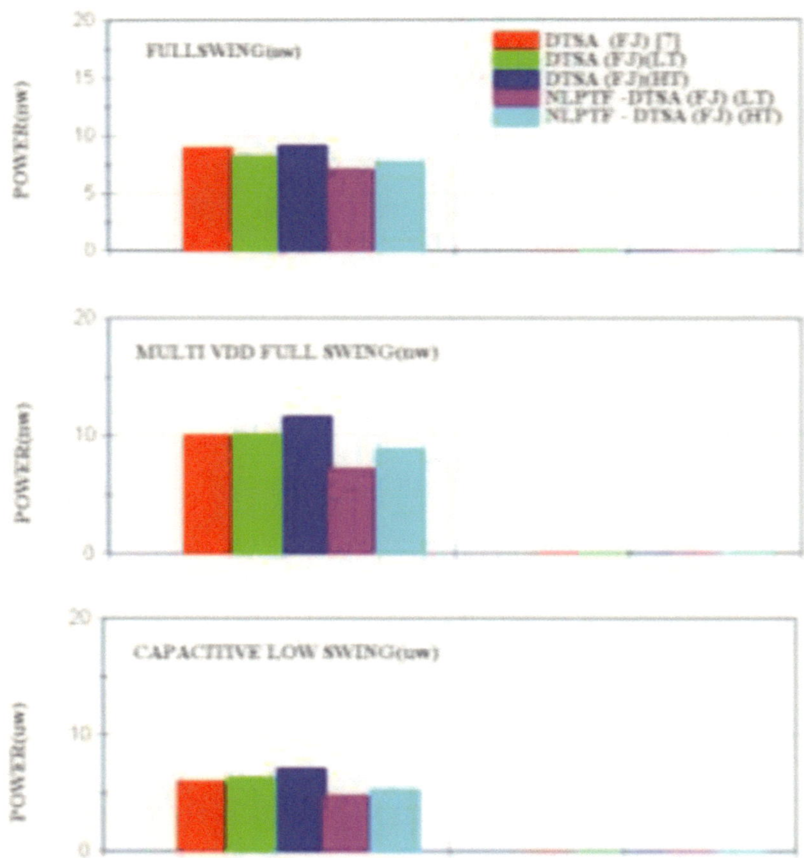

Figure 11.
Static Power comparison of DTSA modules.

Work	Module name	Traffic mode	Data rate GB/S (data rate improvement %)	Link power (W)	Latency (10 mm of interconnect) single/ five stage operation
[7]	DTSA	—	5.0 (80%)	0.8	300/1500
Proposed work	DTSA	LT	4.9 (78.4%)	0.98	345/1725
Proposed work	DTSA	HT	4.0 (64%)	1.32	487/2435
Proposed work	NLPTF-DTSA	Average	5.92 (94.72%)	0.51	497/2485

Table 3.
The overall transceiver performance comparison.

The conservative method has achieved latency of 300/1500 ps under various stage operation (one and Five). The latency result of the MAS work is slightly increased to 440/2200 ps.

Though the latency results are high, still it is encouraging because we added traffic generator and Traffic Estimator.

5. Conclusion

The proposed work is summarized into Three stages namely selection, analysis, design and performance comparison. In the first stage, among various SA's suitable SA'S (DTSA and NLPTF-DTSA) is selected for MAS process. In the second stage, Traffic action takes place according to both DTSA with Transceiver and NLPTF-DTSA Transceiver.

On the Final stage, we compared both above circuitry and concluded under various traffic conditions NLPTF-DTSA is suitable. The result of the complete transceiver circuit (NLPTF-DTSA) under average traffic mode is attained as 5.92 Gb/s data rate, 0.62 W link power and latency of 497 ps/2485 ps for single/five stage operation. When compared with conservative methods, results in MAS design show performance enhancement of 94.72% in data rate and 0.51 W reductions in link power. Though latency of MAS design is high, it is acceptable because of the addition of TG and TE, which is not present in conventional NoC architecture. In future we will improve the performance of NoC Architecture with respect to latency.

Author details

Erulappan Sakthivel* and Rengaraj Madavan
PSR Engineering College, Anna University, Sivakasi, India

*Address all correspondence to: vlsisakthivel@gmail.com

References

[1] Marculescu R, Bogdan P, "The chip is the network toward a science of Network-on-Chip Design," ELECTRON DES, vol. 2, pp. 371-461, 2007.

[2] Moraes F, Calazans N, Mello A, Muller L, Ost L,"Hermes: an infrastructure for low area overhead packet switching networks on chip," INTEGRATION, vol. 38, pp. 69-93, Oct. 2004.

[3] McKeown N,"The islip scheduling algorithm for input queued switches," IEEE ACM T NETWORK, vol. 7, pp. 118-201, Apr. 1999.

[4] Fang FW, Wong, MDF, Chang YW, "Flip chip routing with unified area io pad assignments for package-board co design," in Conf. IEEE DAC, 2009 pp. 336-339.

[5] Liu YI , Liu G, Yang Y, Li Z, "A novel low swing transceiver for interconnection between NoC routers," in Conf. IEEE DCMT, 2011, pp. 39-44

[6] Larsson P, "Resonance and damping in cmos circuits with on chip decoupling capacitance," IEEE T CIRCUITS I, vol. 45, pp. 849-858, Jul. 1998.

[7] Schinkel D, Mensink E, Klumperink EAM, Tuijl EV, Nauta B, "Low power high speed transceivers for network-on-chip communication," IEEE T VLSI SYST, vol. 17, pp. 12-21, Jan. 2009.

[8] Zhao P, McNeely J, Kaung , Wang N, Wang Z, "Design of sequential elements of the low power clocking system," IEEE T VLSI SYST , vol. 19, pp. 914-918, May. 2011. .

[9] C.-C. Yu. Design of low power double edge triggered flip-flop circuit. In: Proc. 2nd IEEE Conf. Industrial Electronics Applications, 2007, pp. 2054-2057.

[10] V. Stojanovic and V. G. Oklobdzija, "Comparative analysis of master-slave latches and flip-flops for high-performance and low-power systems," IEEE Journal of Solid-State Circuits, vol. 34, no. 4, pp. 536-548, 1999.

[11] Lu Z, Jantsch A, "Traffic configuration for evaluating networks on chips", in Conf. IEEE SoC for Real Time Applications Proceedings, 2005, pp.535-540

[12] Xingwei W , Dingde.J , Zhengzheng X, henhua C, "An accurate method to estimate traffic matrices from link loads for QoS provision," Journal of Communications and Networks, vol. 12, pp. 624-631, Dec. 2010

[13] Kun Wang, Xi an, Changshan Wang, Huaxi Gu, "Quality of service routing algorithm in the torus based network on chip," in Conf. ASICON '09. IEEE, 2009, pp. 952-954.

[14] Junsheng Lv , Beijing, Hainan Liu , Ye M , Yumei Zhou, "An energy recovery D flip flop for low power semi custom ASIC design," in Conf. Micro electronics and Electronics, 2010, pp. 33-36.

[15] Erulappan Sakthivel, Veluchamy Malathi, and Muruganantham Arunraja, "VELAN: Variable Energy Aware Sense Amplifier Link for Asynchronous Network on Chip," Circuits and Systems, vol. 7, pp. 128-144, 2016.

[16] Schinkel D, Mensink E, Klumperink EAM, Tuijl EV, Nauta B," A 3-Gb/s/ch transceiver for 10 mm uninterrupted RC limited global on chip interconnects," IEEE J SOLID-ST CIRC , vol. 41, pp. 297-306., Jan. 2006.

[17] Synopsys, Inc., Mountain View, CA [Online]. Available: http://www.synopsys.com

[18] Hwang YT, Lin, Fa J , Sheu MH, "Low power pulse triggered flip flop design with conditional pulse enhancement scheme," IEEE T VLSI SYST , vol. 20, pp. 361-366, Feb. 2012.

[19] Qiaoyan Yu, Paul Ampad, "A flexible and parallel simulator for networks on chip with error control," IEEE T COMPUT AID D , vol. 29, pp. 103-116, Jan. 2010.

Section 2

Network Architecture and Design

Chapter 3

Network-on-Chip Topologies: Potentials, Technical Challenges, Recent Advances and Research Direction

Isiaka A. Alimi, Romil K. Patel, Oluyomi Aboderin, Abdelgader M. Abdalla, Ramoni A. Gbadamosi, Nelson J. Muga, Armando N. Pinto and António L. Teixeira

Abstract

Integration technology advancement has impacted the System-on-Chip (SoC) in which heterogeneous cores are supported on a single chip. Based on the huge amount of supported heterogeneous cores, efficient communication between the associated processors has to be considered at all levels of the system design to ensure global interconnection. This can be achieved through a design-friendly, flexible, scalable, and high-performance interconnection architecture. It is noteworthy that the interconnections between multiple cores on a chip present a considerable influence on the performance and communication of the chip design regarding the throughput, end-to-end delay, and packets loss ratio. Although hierarchical architectures have addressed the majority of the associated challenges of the traditional interconnection techniques, the main limiting factor is scalability. Network-on-Chip (NoC) has been presented as a scalable and well-structured alternative solution that is capable of addressing communication issues in the on-chip systems. In this context, several NoC topologies have been presented to support various routing techniques and attend to different chip architectural requirements. This book chapter reviews some of the existing NoC topologies and their associated characteristics. Also, application mapping algorithms and some key challenges of NoC are considered.

Keywords: Application mapping, interconnection, latency, scalability, system-on-chip (SoC), topology, network-on-chip (NoC), NoC design, on-chip network

1. Introduction

Distributed or parallel systems have been the main approaches of attending to the demand for applications in which huge computations are required. In these systems, a number of processing elements are connected by means of an interconnection network. It is noteworthy that the conventional off-chip architecture in which different processing elements are interconnected is unsuitable for satisfying the requirements regarding scalability, high-throughput, and low-power consumption. This is owing to the increase in delays and hardware complexities of the existing computer chips [1, 2].

The rising computational requirements of modern applications and services have shifted research attention to improvements in semiconductor-based technology. This has resulted in the evolution of on-chip networks. Based on the salient features such as low footprint and power consumption, on-chip-based networks are gaining significant attention compared to the off-chip counterpart. Instances of such on-chip-based architectures are System-on-chip (SoC) and multiprocessor-SoC (MPSoC) [3].

Furthermore, a variety of the on-chip network components are connected using various interconnection networks such as shared bus and bus. Communication among devices in bus-based topology normally occurred on the bus links, while wire collections are employed in the shared bus topology. Comparatively, the shared bus architecture offers a low-cost solution and simple control features. These benefits make the shared bus network a preferred architecture for communication among the on-chip integrated processing units [1, 4].

In addition, it is challenging for the bus-based SoC to meet the requirements of different applications due to the growing increase in the number of Intellectual Property (IP) cores as well as other on-chip resources [2]. Besides, diverse communication requirements are demanded by hybrid processing network elements. The limitations are mainly owing to the bus-based interconnection architecture's inability to offer the required latency, scalability, bandwidth, and power consumption for supporting the huge number of on-chip resources. So, the requirements are challenging for the SoC architectures to satisfy. The on-chip communication bottleneck can be effectively addressed with the implementation of network-on-chip (NoC). Apart from being able to attend to the bus architectural delays and congestion-related problems, the communication requirements can be met with lower power consumption and higher efficiency by the NoC compared to bus-based SoC [1, 4, 5].

This chapter presents a comprehensive overview of the evolution of NoC architectures and their associated features. In this regard, it focuses on a number of major and promising on-chip research areas such as topology, routing, and switching. Also, different application mapping algorithms for enhancing the on-chip performance are presented. Moreover, open problems in on-chip communication design and implementation are discussed. This chapter is organized as follows. Section 2 presents a comprehensive discussion on the SoC with a main section focus on the on-chip interconnect evolution. In Section 4, an in-depth discussion on typical NoC architectural components is presented. Besides, NoC-based application task representation and application mapping are discussed. Section 5 focuses on various NoC topology performance assessment and metrics with some models. Also, Section 6 discusses the related challenges of on-chip schemes and concluding remarks are given in Section 8.

2. System-on-chip

This section focuses on the on-chip interconnect evolution. Conceptually, SoC comprises a circuit that is embedded on a small coin-sized chip and integrated with a microprocessor or microcontroller. Moreover, in the SoC, a single chip is partitioned into functional tiles and an interconnection (communication) network. Depending on the application, the SoC design typically contains storage devices, RAM/ROM memory blocks, central processing unit (CPU), input/output ports, and peripheral interfaces such as timers, inter-integrated circuit (I^2C), universal asynchronous receiver/transmitter (UART), graphics processing unit (GPU), controller area network (CAN), serial peripheral interface (SPI), and so on. Also, based on the requirement, a floating-point unit or analog or digital signal processing system can as well be included.

Furthermore, with innovative integration technology, a huge amount of heterogeneous cores can be supported on a single chip [6, 7]. So, in designing such a chip, the interconnections between multiple cores have a considerable influence on the system performance and communication. For instance, apart from being in charge of all memory transactions, on-chip communication architecture is also responsible for I/O traffic management. Besides, it offers a reliable channel for data sharing between processors. Therefore, high-performance and scalable on-chip communications are highly imperative in the SoC design [8]. However, efficient communication among the on-chip components is challenging [1, 2, 4, 8]. **Figure 1** illustrates an on-chip interconnect architectural evolution.

2.1 Bus architecture

As aforementioned, the traditional on-chip bus-based interconnect techniques are widely used partly due to protocol and architectural design simplicities. Other salient features of bus topology are low area overhead and predictable latency. For instance, bus-based interconnect is mainly suitable in a scenario with a small amount of on-chip components. Besides, it is not only efficient regarding power but also based on the silicon cost [1, 4, 8].

Furthermore, as illustrated in **Figure 1(a)**, a single control/data bus is used in this architecture to support multiple component interactions. This helps in ensuring a simple master–slave connection. Moreover, resource contention can also occur in bus-based architecture when multiple masters are required to communicate with the same slave. In such a situation, arbitration is demanded to ensure effective communication. Therefore, in a large SoC, the bus-based architecture scalability is challenging [8].

2.2 Crossbar (matrix switch fabric) architecture

As discussed in subSection 2.1, when multiple master–slave data transactions are to be supported, a single shared bus architecture is not a suitable solution due to the associated latency. This is as a result of the arbitration that is demanded between the master interfaces on the shared medium. Consequently, crossbar topology can be employed to address the scalability issue of on-chip interconnect. As illustrated in **Figure 1(b)**, a crossbar architecture consists of a matrix switch fabric. To facilitate multiple communications, this matrix connects the entire inputs with the entire outputs. Based on the advantages of the crossbar topology, it has been adopted by the SoC designer by merging multiple shared busses to achieve a connection matrix

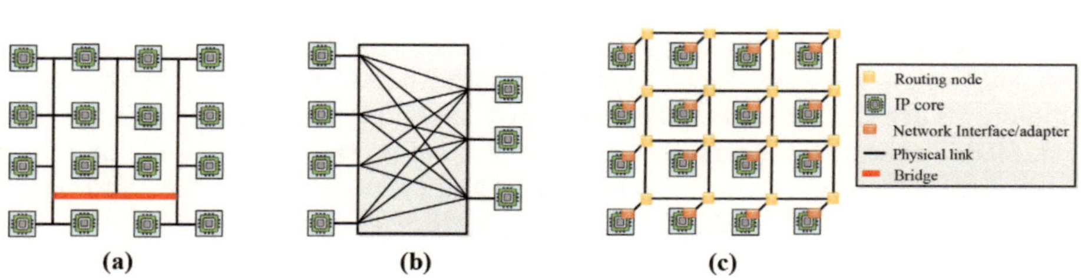

Figure 1.
On-chip interconnect architectural evolution: (a) bus, (b) crossbar, and (c) 4 × 4 mesh-based NoC.

with an all-input-all-output [8, 9]. Nonetheless, the design complexity of a crossbar-based architecture is high owing to the layout of wire that is complex [8].

2.3 Network-on-chip architecture

NoC offers an alternative and modular nature platform with high scalability. Besides, it supports efficient on-chip communication that facilitates NoC-based multiprocessor architectures with high functional diversity and structural complexity [3]. This makes it a *de facto* on-chip communication standard for highly integrated SoC architectures. Besides, it supports parallel (multiple concurrent) communications by enabling pipelining irrespective of the network size. Also, rather than establishing a connection between all IP blocks, in the NoC, a network is created within the chip. This enables each IP to function as a network node. For instance, to ensure effective communication, a network of routers is employed to connect the associated huge cores. In this regard, the bus in SoC has been replaced by a network of routers that controls the communication process among nodes in the established network. Based on this, NoC presents a number of characteristics such as low-latency, high-bandwidth, and scalability [10, 11].

In the NoC, the interconnections are suitably organized to form appropriate topologies. Moreover, communication in it normally transpires between IP cores and in accordance with the employed topology. Also, this can be achieved using asynchronous or synchronous modes [11]. So, with these topologies, certain routing techniques can be employed for packet routing between nodes [10]. As depicted in **Figure 1(c)**, components such as routers and channels (interconnection links) are required for packet routing [11]. It is noteworthy that some routing techniques are specially intended for NoC. As a result, they are well-designed to be deadlock[1]-free [10].

Furthermore, **Figure 1(c)** illustrates a 4×4 mesh-based NoC architecture with a number of processing cores/processing elements (PEs) connected via regular-sized wires and routers. The PEs can be components such as application-specific integrated circuit (ASIC) block and microprocessor [12]. Moreover, different types of cores such as the manager, regular, and spare can be employed. Also, depending on the application, these cores can be homogeneous and heterogeneous. The regular cores normally execute the task of a specified application, the spare cores are additional cores that can be employed in case of failure of either regular or manager core, and the manager cores are used to track and manage all processing cores. Besides, when a processing core fails, the manager core performs the task migration [13].

Moreover, a network interface (NI) is placed at the edge of each PE and on-chip interfaces such as high-definition multimedia interface (HDMI), I^2C, USB, and UART are supported. The routers are employed to packetize the generated data by the PE. The NI is subsequently connected to a router that buffers the packets from the PE or other connected routers [12]. In this regard, the NI connects the NoC routers and hardware IP blocks. Consequently, NI facilitates the modular property and ensures seamless communication between different IP with related housekeeping operations, irrespective of their communication protocol [8].

In addition, it is noteworthy that the experienced design problems in the NoC architectures are similar to the ones in the bus-based architectures. In this context, to establish a communication fabric capable of meeting the requirements of a particular application, a trade-off between reliability, power, area, cost, and performance is demanded [12]. For instance, asymptotic cost functions for a shared

---

[1] A deadlock happens in the NoCs when one or more packets remain blocked for an indefinite time. It can be addressed either by imposing routing restrictions or by employing additional hardware resources.

	Power Dissipation	Operation Frequency	Total area
Shared bus	$\mathcal{O}(n\sqrt{n})$	$\mathcal{O}(\frac{1}{n^2})$	$\mathcal{O}(n^3\sqrt{n})$
Segmented bus	$\mathcal{O}(n\sqrt{n})$	$\mathcal{O}(\frac{1}{n})$	$\mathcal{O}(n^2\sqrt{n})$
Point-to-Point	$\mathcal{O}(n\sqrt{n})$	$\mathcal{O}(\frac{1}{n})$	$\mathcal{O}(n^2\sqrt{n})$
NoC (Mesh)	$\mathcal{O}(n)$	$\mathcal{O}(1)$	$\mathcal{O}(n)$

Table 1.
Asymptotic cost functions for interconnection architectures.

bus, segmented bus, point-to-point, and NoC interconnect with n system modules are presented in **Table 1**. This shows that with a growing n, NoC architecture dissipates less power, requires less wiring area, and offers excellent operating frequency, making it a scalable and attractive architecture [14].

3. Advanced on-chip bus architectures

Since there are various IPs with distinct standard interfaces from different providers, the chip designers have to adapt to connect through common standard or in-house interfaces. Consequently, a flexible and open standard for IP core interface is essential for practical on-chip interconnection design and SoC integration. This can be achieved by employing standard interface protocols that offer reusable profiles that can support diverse on-chip interconnection design and SoC integration. Also, the operation of an on-chip interconnection depends on the bus architecture efficiency. So, bus architecture with additional data transfer cycle, faster clock speed, enhanced throughput, and width is highly attractive for a reduced time-to-market, low cost, and efficient SoC. This section presents an overview of standard on-chip bus structures and protocols such as ARM Advanced Microcontroller Bus Architecture (AMBA), IBM CoreConnect, and Altera Avalon.

3.1 AMBA-based bus protocol architecture

The AMBA bus protocols are the ARM interconnect specifications for on-chip communication between a number of functional blocks. In these designs, one or more microprocessors/microcontrollers can be integrated on a single chip with various other components and peripherals. **Figure 2** depicts a traditional AMBA 2.0 based SoC design with Advanced System Bus (ASB) or Advanced High performance (AHB) protocols and an Advanced Peripheral Bus (APB) protocol for high bandwidth and low bandwidth peripheral interconnections, respectively [15].

Moreover, to scaling up connectivity and address the limitations regarding the number of IPs that can be effectively supported by the AHB/ASB protocols, AMBA 3 presents Advanced Extensible Interface (AXI) interconnect for point-to-point

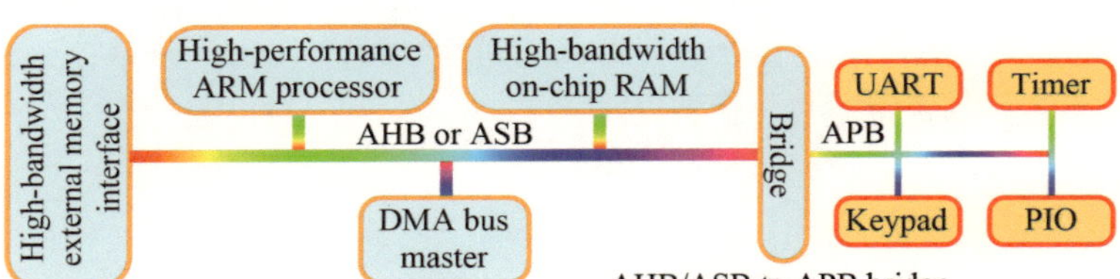

Figure 2.
A typical AMBA based SoC design. PIO: Peripheral I/O, UART: Universal asynchronous receiver/transmitter,

connectivity protocol. Some of the main features of the AXI protocol are the capability to issue multiple outstanding transactions, unaligned data transfers with byte strobes, separate control/address and data phases, simultaneous read and write data channels to guarantee low-cost Direct Memory Access (DMA), and out-of-order data capability. **Figure 3** shows AXI interconnect enabling IPs communication with a master–slave protocol. It is noteworthy that the interconnect such as a switch design, a convention crossbar, or an off-the-shelve NoC capable of supporting multiple AXI masters and slaves can be employed. Also, an array of peripherals supported on an APB bus are connected through an AXI to APB bridge [15, 16].

Furthermore, the emergent of portable mobile devices such as smartphones in which SoCs are equipped with dual/quad/octa-core processors and shared integrated caches demand hardware managed coherency within the memory subsystem, resulting in the development of AXI Coherency Protocol Extension (ACE) in the AMBA 4. Also, with the current trend towards heterogeneous computing for improving the performance of data center, parallel, and High-Performance Computing (HPC) applications, numerous heterogeneous computing elements, processor cores, and IO subsystems are demanded. To support the requirements, the Coherent Hub Interconnect (CHI) protocol was presented in the AMBA 5 protocol to improve the AXI/ACE protocol design. For instance, for better scalability, the associated signal-based protocol in the AXI/ACE was changed to a packet-based layered protocol in the CHI. Some of the supported features are Cache stashing, Cache de-allocation transactions, atomic transactions, and Persistent Cache Maintenance Operations (CMO). Other AMBA specifications with additional efficient translation services and higher performance are Distributed Translation Interface (DTI) and Local Translation Interface (LTI) protocols [16, 17].

3.2 WishBone bus protocol architecture

WishBone interconnect primarily focuses on design reuse to address integration problems by establishing a general-purpose interface between IP cores. This helps in improving the system's portability and reliability. This interconnect comprises two interfaces which are master and slave. The IPs are master interfaces that can initiate bus cycles. Also, the slave interfaces accept the initiated bus cycles. Besides, its hardware implementations are compatible with varieties of interconnection such as dataflow, crossbar-switch, shared bus, and point-to-point [16].

Furthermore, WISHBONE specifies a single, simple, logical, synchronous MASTER/SLAVE bus and IP core interfaces that demand very few logic gates. Also, it supports some standard data transfer protocols such as BLOCK READ/WRITE cycles, SINGLE READ/WRITE cycles, and read-modify-write (RMW) cycles.

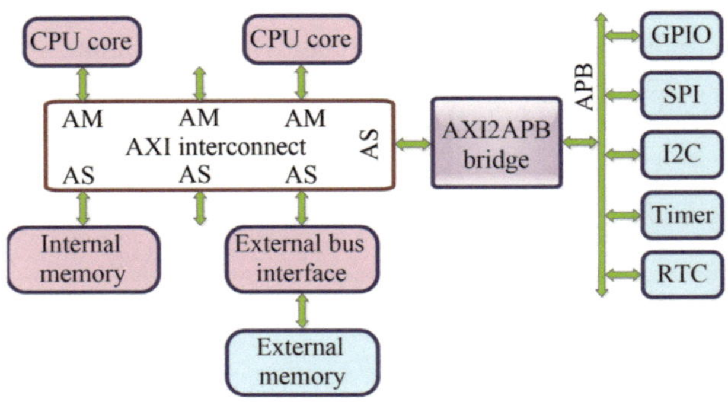

Figure 3.
An AXI interconnect. GPIO: General purpose input/output.

Moreover, the related flow control and communication among the cores are facilitated by the handshake mechanism. Besides, its multiprocessing capabilities enable a broad range of SoC configurations [17, 18].

3.3 Open core protocol

Open core protocol (OCP) is an open standard, non-proprietary, core-centric protocol for attending to the requirements of IP core system-level integration. Also, it defines a clocked system that offers unidirectional data transfer that helps in simplified core integration, implementation, and timing analysis. Moreover, based on its high configurability and flexibility, it supports independent IP cores design and facilitates IP reuse. Based on this, it enhances and ensures IP modularity without the need for redesign. Besides, all test/debug and sideband signals are offered by the OCP for a number of functions, such as protections or interrupts. Also, some of its features and signals are optional. This helps the users in choosing the configuration that best suits their IP cores [17].

A typical OCP operation across an on-chip interconnect is shown in **Figure 4**. In this configuration, an OCP master/slave element is integrated into IP cores. The implementation comprises a request and a response channel. The master IP core issues read command that causes a transfer on the request channel. On the response channel, the slave IP core responds to the master IP core. Also, some supported extensions by the OCP protocol are the transfer of a burst of data, data handshake extensions, out-of-order responses, and test control extension [16].

3.4 IBM CoreConnect architecture

IBM CoreConnect™ architecture is another open on-chip bus structure that offers the framework for efficient realization of complex SoC designs. As illustrated in **Figure 5**, it has three separate busses for interconnecting cores, custom logic, and

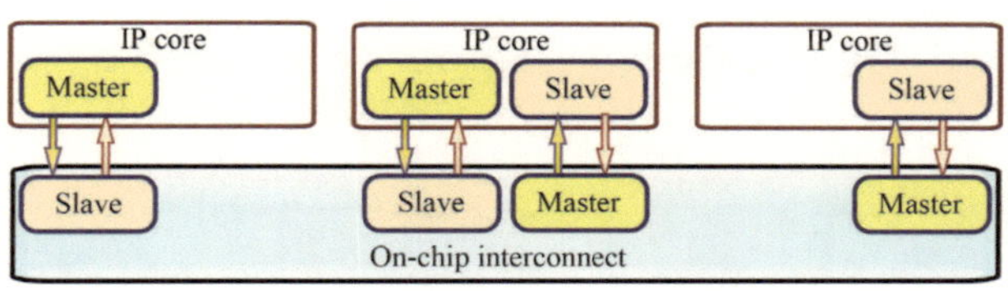

Figure 4.
Typical block diagram of OCP.

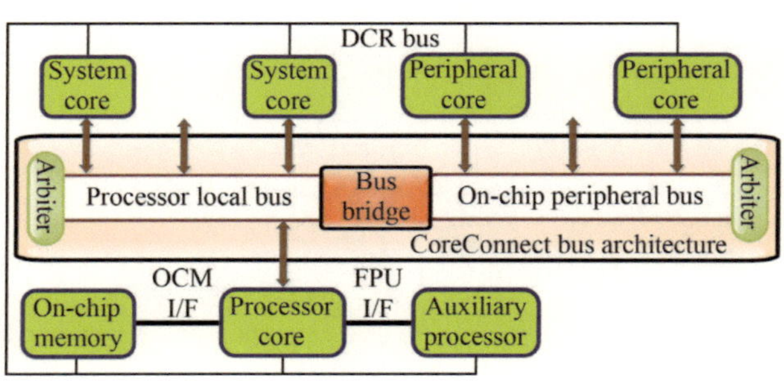

Figure 5.
Typical block diagram of CoreConnect.

library macros are On-Chip Peripheral Bus (OPB), Processor Local Bus (PLB), and Device Control Register Bus (DCR). The architecture can be employed for different customer-specific and application-specific SoC designs in high-performance embedded applications, storage, networking, wired/wireless communications, and low-power pervasive applications. In this context, high-performance peripherals can be connected to the low latency, high bandwidth PLB. Also, device-paced peripheral cores are normally connected to the OPB. This helps in reducing traffic on the PLB and consequently, enhancing the system performance. Also, a relatively low-speed data path is offered by the DCR bus for control, initialization, and status information [16, 18].

3.5 Altera Avalon architecture

Avalon presents a simple bus architecture for the connection of on-chip peripherals and processors to a system-on-a-programmable chip (SOPC). Also, the offered interface defines a port for connecting the master and slave components and the timing for the components' communication. Besides, it supports multiple masters that present construction flexibility for SOPC systems. The slave-side arbitration is used in the masters and slaves interaction. So, if multiple masters try to access the same slave simultaneously, slave-side arbitration logic controls the master that gains access to the slave to complete the requested transactions. **Figure 6** illustrates a typical block diagram of an Avalon bus module with a collection of connected peripherals [17, 19].

Moreover, the Avalon bus module comprises data, address, and control signals, and arbitration logic that are needed for connecting the peripheral components. Also, its operation comprises address decoding for the selection of peripheral and wait-state generation for supporting slow peripherals. Furthermore, apart from the simplicity and optimized resource utilization, the bus also offers synchronous operation and dynamic sizing. Also, different transactions can be realized between a master and slave peripheral. Likewise, different advanced features such as multiple bus masters, streaming peripherals, and latency-aware peripherals are supported. Based on this, during a single bus transaction, multiple data units can be conveyed between peripherals [18, 19]. **Table 2** compares different SoC busses.

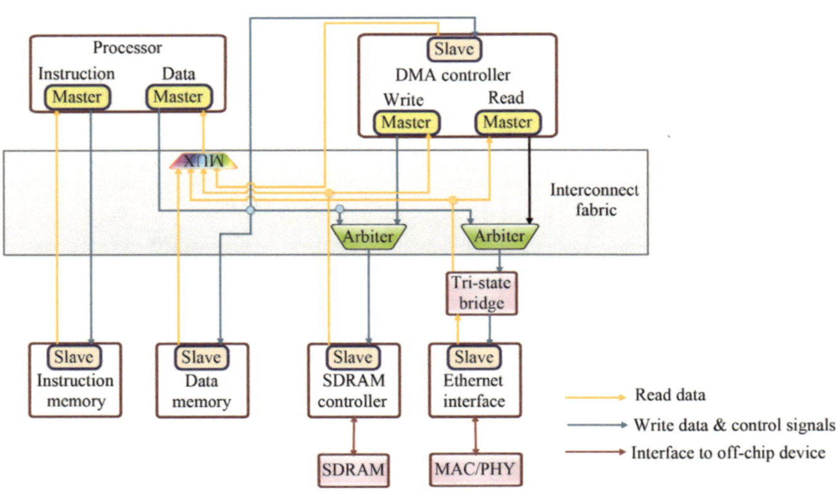

Figure 6.
Typical block diagram of Avalon.

DOI: http://dx.doi.org/10.5772/intechopen.97262

Protocol	Bus Owner	Bus Topology	Arbitration	Bus Width (bits)		Transfers
				Data	Address	
AXI	ARM	Bus-Matrix & Hierarchical	Static Priority, TDMA, Lottery, Round-Robin, Token-passing and CDMA	8, 16, 32, 64, 128, 256, 512, or 1024	32	Handshaking, Split, Pipelined and Burst
Wishbone	OpenCores.org & Silicore Cooperation	Point-to-Point, Crossbar Connection, Shared & Data-flow Interconnection	Static Priority, TDMA, Lottery, Round-Robin, Token-passing and CDMA.	8,16,32,64	1–64	Handshaking & Burst
OCP	OCP Int. Partnership	Interconnect Topology	Vary depending on logic on the bus side of OCP.	Configurable	Configurable	Handshaking, Split, Pipelined & Burst
Avalon	Altera	Point-to-Point, Pipelined, Multiplexed	Static Priority, TDMA, CDMA, Round-Robin, Lottery, Token-passing	1–128	1–32	Pipelined and Burst
CoreConnect	IBM	Hierarchical	Static Priority	PLB (32, 64, 128 or 256); OCB (8, 16 or 32) and DCR (32)	PLB and OPB (32) and DCR (10)	Handshaking, Split, Pipelined and Burst

Table 2.
Comparison of SoC busses.

4. Network-on-chip components

As aforementioned, a network of routers is employed in the NoC for controlling the communication process among nodes. Several topologies along with different routing algorithms have been presented for NoC architectures. It is noteworthy that the network topology selection is a primary step in the network design. Besides, the flow-control techniques and routing strategy depend greatly on the topology. This section focuses on typical architectural components such as network topology, switching, and routing algorithm. Besides, task representation and application mapping are presented.

4.1 Network-on-chip topologies

The NoC topology denotes the physical organization of its architecture and it signifies a key design criterion. In this context, the means by which its elements are interconnected are characterized. The NoCs have been considered as regular tile-based topologies that are appropriate for connecting homogeneous cores. Besides, much attention has been given to custom-based, domain-specific irregular topologies to support heterogeneous cores with diverse size, functionality, and communication requirements [3]. Some of these topologies are discussed in this subsection.

4.1.1 Regular topologies

In regular topologies, the power consumption and network area scalability with an increase in the size can be predicted. It should be noted that regular network topologies are usually adapted for the majority NoCs [20]. This subsection focuses on the most popular regular topologies along with their advantages and drawbacks.

Ring Toplogy:

Ring topology is one of the widely employed NoC topologies. In this topology, a single wire is used to connect each node. Consequently, irrespective of the ring size, each of the nodes has neighboring nodes as depicted in **Figure 7(a)**. Based on this, in the ring topology, the degree[2] of each node is two. This implies a corresponding available bandwidth to every node. Although deployment and troubleshooting are comparatively easier, the main drawback of the ring topology is that its diameter increases with an increase in the number of nodes. So, besides the fact that network expansion degrades the performance (scalability issue), ring topology is also prone to a single point of failure (poor path diversity) [1, 21].

Octagon Topology: Another prevalent NoC topology is the octagon. A typical octagon topology comprises eight (8) nodes and twelve (12) bidirectional links. Also, just like the ring topology, each node is connected to the preceding and succeeding nodes. So, between a node pair, there are two-hop communications. Also, to route a packet between the network, a simple shortest-path routing can be employed. Besides, compared with a shared bus topology, higher aggregate throughput can be achieved. Furthermore, the architecture can be connected to support bigger designs, resulting in better scalability.

Star Topology: The star topology in which the entire nodes are connected to a central node is shown in **Figure 7(c)**. Assume an N nodes with $N-1$ connected nodes to the central node. In this architecture, the central node has an $N-1$ degree

[2] Router degree is a parameter that specifies the number of on-chip components and neighboring routers that it is connected to. It is noteworthy that the router microarchitecture complexity increases with an increase in its degree.

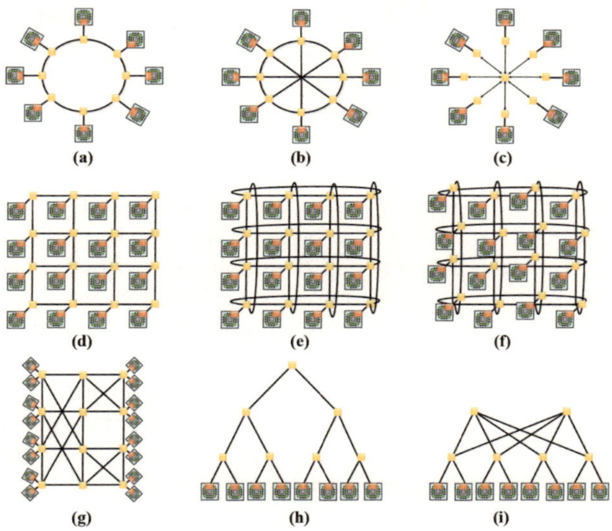

Figure 7.
NoC topologies: (a) ring, (b) octagon (c) star, (d) 4 × 4 mesh, (e) 4 × 4 torus, (f) 4 × 4 folded torus, (g) butterfly, (h) binary tree, (i) fat tree.

while others have a degree of 1. Therefore, regardless of its size, the star topology diameter is 2. In this regard, its main benefit is the offered simplicity and the presented minimum hop count of two due to the associated small diameter [21]. Although the nodes are separated and free of the potential impact from the failed nodes, the central node failure can result in the entire network failure. Furthermore, as the diameter of the central node increases with the number of nodes, a communication bottleneck can take place in the central node [1].

Mesh Topology: The mesh architecture is the widely employed interconnection topology. A typical 4 × 4 mesh topology with 16 functional IP blocks is illustrated in **Figure 7(d)**. Besides the router at the edges, each router in the mesh topology is connected to one computation resource and four neighboring routers through communication channels. With mesh topology, a huge number of IP cores can be incorporated in a regular-shape structure [4]. So, this topology offers an attractive solution for path diversity and scalability. Likewise, this topology can tolerate link failure due to multiple paths that connect a pair of nodes [21]. Nevertheless, one of the main challenges of this topology is that its diameter increases significantly with the number of nodes. This is owing to irregularity in the degree [1]. For instance, the degree of corner, edge, and inner nodes are 2, 3, and 4, respectively [21]. Besides, the associated bandwidth often varies from one node to another, with corner and edge nodes having lesser bandwidth [1, 21].

Torus Topology: A typical torus topology is depicted in **Figure 7(e)**. The architecture is very similar to a mesh topology. However, mesh topology offers a reduced diameter. Consequently, the challenge of diameter increase of mesh topology with the network size is addressed by the torus topology. This is achieved through the addition of direct connections between the end nodes that are in the same column or row [21]. For instance, in the torus topology, wrap-around channels are employed for the connection of the edge routers to those at the opposite edge, resulting in a better bisection bandwidth and reduced average number of hops. However, considerable latency is incurred by the torus topology due to the employed lengthy wrap-around connections [1, 4].

In addition, an alternative to the torus is the folded torus topology. The folded torus topology offers a shorter link length, resulting in reduced implementation area and traverse time for the packet between the interconnected links. Compared with the torus, folded torus offers more path diversity, making it more fault-tolerant. Besides, as aforementioned, torus topology helps in reducing the associated mesh latency. Nevertheless, its long wrap-around links can cause undue delay. This challenge can be addressed by folding the torus as depicted in **Figure 7(f)**.

Butterfly Topology: A typical butterfly topology is illustrated in **Figure 7(g)**. It offers a fixed hop distance between any source to the destination node pair and the router degree is 2, resulting in low-cost routers. Owing to the single path that exists from the source to the destination node, the topology lacks path diversity, resulting in low link fault tolerance and low bandwidth. Besides, this topology normally entails lengthy wires and the related complex wire layout can lead to more energy consumption [1, 21].

Binary Tree Topology: The binary tree topology consists of a top (root) node and bottom (leaves) nodes illustrated in **Figure 7(h)**. In this configuration, besides the root node, each of the others has two offspring nodes. Also, besides the root node that has no parent, each of the other nodes has its parent and children directly above and beneath itself, respectively. The nodes in this topology have access to broader network resources and it is supported by several vendors. However, its bottleneck is the root node, whose failure can bring about the entire network failure. Also, with an increase in the tree length, network configuration becomes more intricate.

Fat Tree Topology: The concept of fat tree topology is based on using intermediate routers as forwarding routers and connecting the leave routers to the clients as illustrated in **Figure 7(i)**. Although this topology offers excellent path diversity and better bandwidth, the router to clients ratio is extremely high and the wiring layout is complex. Therefore, a number of routers should be integrated to connect with fewer clients [1].

Cube-Based Topology: There are a number of cube-based topologies that have been designed. One such appropriate architecture is a hypercube topology. However, its main disadvantage is that there are restrictions in its network size because of the degree limitation. To address the limitation, various variations such as folded hypercube, dual cube, crossed cube, cube-connected cycles, hierarchical cube, and metacube have been presented. A number of these topologies are depicted in **Figure 8** and are mainly focusing on reducing the associated node degree and/or minimizing the network diameter while the diameter is kept as small as possible [10, 21, 22].

In a folded hypercube, each node is connected to the farthest distinct node. Based on this, there is a considerable reduction in its diameter compared with the hypercube topology. However, this is at the expense of additional links. Furthermore, a crossed cube can be realized through the transposition of some edges in the hypercube. This helps in the diameter reduction without causing additional link complexity. In a reduced-hypercube, to minimize node degree, the edges are reduced from an n-dimensional hypercube [22]. An (n, n) hierarchical cubic network consists of n cluster and each of the clusters has n-cube. Furthermore, a hierarchical hypercube is a dual cube structure. This topology comprises two classes

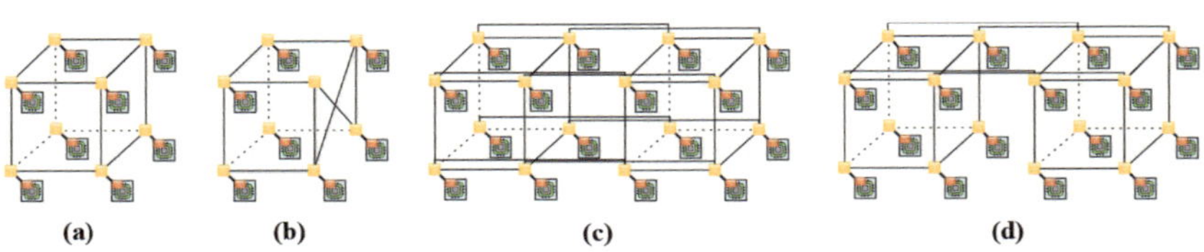

Figure 8.
Cube-based topologies: (a) cube, (b) crossed cube, (c) hypercube and (d) reduced hypercube.

(0 and 1) with clusters. Also, each of the clusters contains $2m$ nodes. Likewise, in an m-dual cube, the binary address of each node is $1 + 2m$ bit long. Similarly, cube-connected cycles offer a hypercube implementation with virtual nodes. In this topology, rather than a single node, each virtual node is a circle with three ports. Also, a metacube topology is a two-level hypercube architecture. It is a symmetric network with a short diameter and small node degree. Structurally, this multi-class topology is an extended form of the dual cube [21].

4.1.2 Irregular topologies

Irregular topologies are based on the integration of various forms, usually regular structures, in different fashions. In this regard, a hybrid, hierarchical, or asymmetric approach can be adopted. Moreover, irregular topologies aims at increasing the available bandwidth compared with the traditional shared busses. Besides, compared with the regular topologies, it helps in reducing the distance among nodes [12]. Also, irregular topologies typically scale nonlinearly with area and power. They are usually based on the concept of clustering and adapted for specific applications [20]. **Figure 9** illustrates some irregular topologies such as reduced (optimized) mesh (**Figure 9(a)**- i and ii) and cluster-based hybrid (mesh + ring- **Figure 9(b)**).

In addition, apart from the classification discussed in subSection 4.1.2, NoC topology can also be categorized as direct[3] and indirect[4] topologies. For instance, ring, bus, mesh, and torus topologies are direct topology. On the other hand, a clos, butterfly, benes, and fat-tree are good instances of indirect topology [11, 12].

4.2 Network-on-chip routing

With suitable topology, a network will be established among the on-chip IPs to ensure effective communication. The communication can be achieved using appropriate algorithms for routing the packets from the source to the destination nodes. In this context, routing algorithms control efficient and correct packet routing as they traverse through the nodes. As aforementioned, starvation[5]-free and deadlock-free routing algorithms are of utmost importance in NoC [10]. Furthermore, the routing algorithm can be selected based on a number of interrelated features,

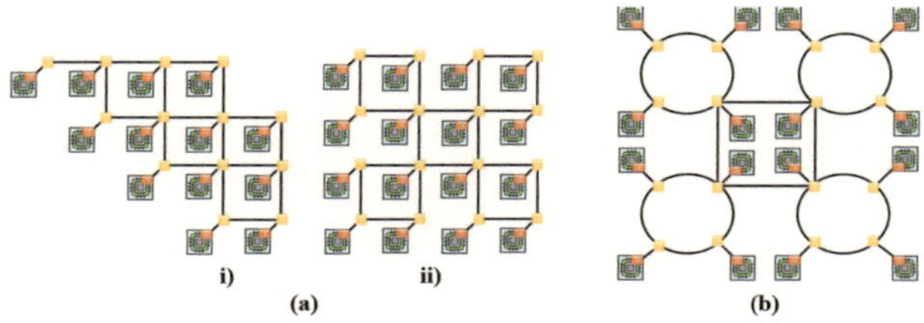

Figure 9.
Irregular topologies: (a) reduced mesh structures and (b) cluster-based hybrid.

[3] In this topology, there is a direct connection of each router to at least a core.

[4] In this topology, some of the employed routers are not directly connected to any of the cores.

[5] Starvation usually occurs in NoCs when specified priority rules are employed for routing, mainly in favor of the high priority packets, making low priority packets wandering in the network. It can be attended to by reserving some resources for the low priority packets and adopting fair routing algorithms.

resulting in trade-offs between related metrics such as the power consumption, packet latency, and footprint that determine the routing algorithm quality. For instance, when the routing circuit is kept simple, the required power for routing can be minimized, and consequently, the power consumption can be reduced. Besides, to increase the performance, the routing tables should be minimized. This will help in ensuring low latency, enhanced robustness low footprint, and effective network utilization [8, 12]. In general, the NoC routing algorithms can be classified based on factors such as the routing path, distance, and decision states. In this context, its NoC routing algorithms can be mainly classified as static and dynamic routing algorithms. Besides, the routing decisions can also be based on distributed, source, minimal, and non-minimal routing algorithms. This subsection focuses on the static and dynamic routing algorithms.

4.2.1 Static routing

Static routing, also known as the oblivious or deterministic routing is the simplest and extensively used routing algorithm in NoCs. It employs fixed (predefined) paths for data transfer between a specific source and destination. Also, the current state of the network is not taken into account in the static routing. So, when making routing decisions, it is oblivious of the load on the links and routers. Static routing requires very little router logic, making its implementation easy. Besides, packets can be split in a scheduled way among several paths between the source and destination. Also, in-order packet delivery can be guaranteed by the static routing in a scenario where just a path is employed. Based on this, the addition of bits to packets at the NI is not required for correct identification and reordering at the destination [12]. Schemes such as random walk routing, directed flood, probabilistic flood, dimension order routing (DOR), destination tag, turn model, *XY*, surrounding *XY*, and pseudo-adaptive *XY*, are examples of static routing algorithms.

XY routing is a distributive deterministic routing algorithm. In this algorithm, the coordinates of the destination address are employed in delivering the packet through a network. The packet is initially routed along the *X* coordinate (horizontal direction) to reach the column. Then, is routed vertically along the *Y* coordinate to its destination [5]. *XY* routing is a preferred algorithm for torus- based and mesh-based topologies [10, 11] and it is deadlock-free. Nevertheless, the associated traffic can be irregular due to the load that is normally created in the middle of the network [10] while *XY* algorithm is not capable of avoiding congested and busy links [5].

4.2.2 Dynamic routing

The routing decisions in adaptive or dynamic routing are based on the existing state of the underlying network. To make routing decisions in this routing scheme, factors like system availability and load condition of the links are considered. Consequently, as the application requirements and traffic conditions change, there can be a corresponding change in the path between the source and destination. Compare with static routing, traffic can be distributed more efficiently across various routers in dynamic routing. Besides, in case of network congestion in certain NoC links, it can exploit alternative paths. In this regard, more traffic can be supported by its topology and the network bandwidth utilization can be maximized. These salient features of the adaptive routing are owing to its ability to exploit the global knowledge of the current traffic state in the optimal path selection [23]. However, this scheme's adaptivity is at the expense of additional resources required for continuous monitoring of the network state to ensure a corresponding dynamic change in the routing paths. This usually presents additional complexity to the router design. Besides, there is a limitation on the effectiveness of adaptive routing due to the constraint on the amount of global knowledge that can be forwarded to each of the routers and also owing to interference [23]. As aforementioned, a static routing scheme is normally employed in scenarios with steady and known traffic requirements, while dynamic routing is primarily applicable to unpredictable and irregular traffic conditions [8, 12]. Schemes such as congestion look-ahead, slack time aware, fully adaptive, minimal adaptive, turnback when possible, turnaround–turnback, odd-even, deflection (hot potato), are examples of adaptive routing algorithms.

In the NoC, to communicate between the source and destination, both adaptive and deterministic routing algorithms can be employed. The odd-even routing algorithm is an adaptive routing and it is a deadlock-free turn model. In this regard, in a grid network, east to south and east to north turns are prohibited in the even columns. Also, north to west and south to west turns are prohibited in the odd columns. So, the odd-even routing algorithm helps in eliminating potential livelock[6] in the system [5, 10].

The deflection routing technique is cost-effective since no buffers are employed. Consequently, the incoming packets received by the routers are not buffered and move simultaneously towards their destinations based on the routing table. However, misrouting can occur when a busy router receives another packet. In a severe situation, the misrouted packets in the network can cause additional misrouting, making each packet to be bouncing around like a *hot potato* across the network [12]. The misrouting can be alleviated considerably with sufficient intervals between the packets [10].

4.3 Network-on-chip switching

The NoC switching scheme denotes the employed switching technique for data control in the routers and specifies the data transfer granularity. Packet switching and circuit switching are the key switching techniques in the NoCs. The switching schemes are illustrated in **Figure 10** and discussed in this subsection.

4.3.1 Circuit switching

The circuit switching is based on the establishment of a reserved physical path (link reservation), consisting of routers and links, between the source and destination before data transmission. Although circuit switching offers low latency transfers due to the full link bandwidth utilization, it wastes the established links when there is no

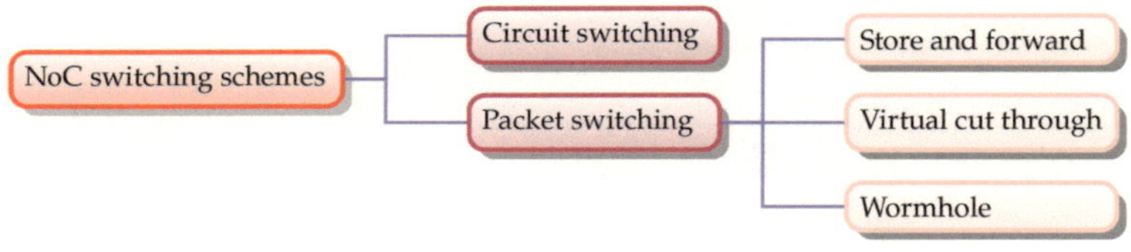

Figure 10.
NoC switching schemes.

[6] A livelock arises in NoCs when a packet bounces around indefinitely between routers without reaching its destination. It is typically associated with adaptive routing and can be addressed by employing uncomplicated priority rules.

data transmission, resulting in scalability issues. Furthermore, to improve network scalability, virtual-circuit switching can be employed. It helps in multiplexing multiple virtual links on a single physical link. Also, the allocated buffers determine the total number of virtual links that the physical link can support [12].

4.3.2 Packet switching

Packet switching is another popular switching mode. Unlike circuit switching in which a path is established prior to the data transmission, in packet switching, there is no need to create a path (no link reservation) before packet transmission. In this context, the transmitted packets follow independent paths (different routes) from the source to the destination. As a result, different delays will be experienced by the packets. Besides, unlike circuit switching in which a start-up waiting time and a fixed minimal latency are normally incurred, in packet switching, a zero start-up time and a variable delay owing to contention are generally incurred. Moreover, due to the contention, Quality of Service[7] (QoS) in packet switching is challenging to guarantee compared with circuit-based switching. Wormhole, virtual cut through as well as store and forward are the extensively employed packet switching schemes [12].

4.4 Network-on-chip application mapping

In supercomputing and parallel computing, application mapping is usually employed for mapping applications that share resources to be in close proximity to minimize the network latency. This is also applicable to the shared bus-based Chip Multi-Processor architectures in which the application mapping should consider the fundamental on-chip interconnect design. Depending on the adopted topology, mapping algorithm implementation helps in the positioning of the IP cores to the NoC tiles. Besides, its performance is highly contingent on the employed routing interface and shared memory architecture [8].

In MPSoC, quite a lot of techniques can be employed for application mapping. Also, the presence of several and diverse MPSoC architectures further complicate the issue. Consequently, in practice, it is advisable to rebuild the mapping approach based on the application-architecture category [8]. Furthermore, as depicted in **Figure 11**, the NoC application mapping algorithms are broadly grouped into static and dynamic based on the assigned task time. In this context, the time at which the tasks are allocated to the IP cores for processing is considered. For instance, in dynamic mapping, the application task clustering, ordering, and assignment to the cores are

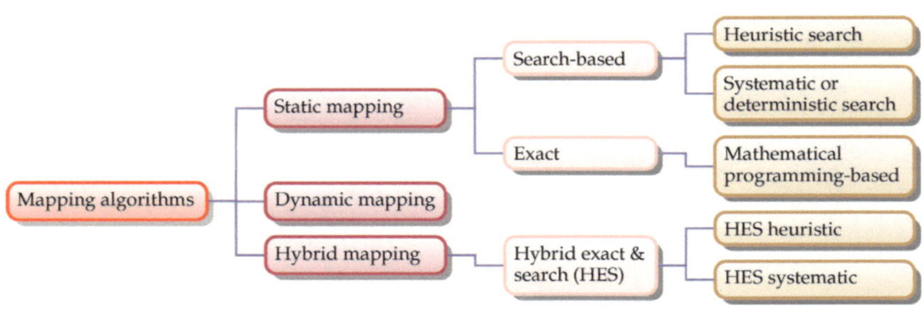

Figure 11.
NoC mapping algorithms classification.

---

[7] Quality of Service implies performance bounds regarding the delay, bandwidth, and jitter; and can be categorized into differentiated service, guaranteed service, and best effort.

implemented in the course of application execution (real-time). Also, dynamic mapping is an efficient solution due to its ability for mapping based on the cores' runtime load. Besides, through the analyzes of the traffic load, the workload can be distributed between the cores to address network congestion. Based on this, the performance bottleneck can be identified at any core. Nonetheless, owing to its related computational complexity (overhead), implementation of the real-time mapping algorithm incurs not only execution delay but also consumes more energy [24, 25].

In static mapping, during the design time, the application task mapping is performed in the off-line. In this context, the mapping is usually finalized prior to the application execution. Since the related application scenarios are known in the design period, optimal or at least near-optimal solutions can be formulated. This makes the static mapping algorithm a good solution for attending to the associated additional communication overhead of the dynamic mapping. Consequently, the related delay and energy consumption can be minimized. However, static mapping can not handle dynamic scenarios that are usually encountered in nature. Furthermore, static mapping is mainly grouped into exact (mathematical based) and search-based algorithms. Search-based mapping algorithms can be further categorized as heuristic and deterministic (systematic) [24, 25].

In addition, hybrid application mapping algorithms have been presented to address the challenges of the aforementioned mapping algorithms by exploiting their advantages. In this regard, hybrid algorithms offer efficient application mapping solutions. Further information on the NoC mapping algorithmsÂ´ classification can be found in [24, 25]. Besides, another promising area is the multiple layer processing core integration into a 3D design. This can considerably help in reducing the power, area, and delay in signal transmission. Based on this, 3D multicore architectures have been considered as a potential solution for future high-performance systems. However, the related high integration density of the 3D presents additional concern regarding the temperature increase. This effect can bring about high-temperature gradients and thermal hot spots that can make the system unreliable and consequently degraded performance. Therefore, the 3D thermal management demands further research attention [24].

5. Topology performance assessment

Some features determine the performance of the NoC-based system and influence the effectiveness of the related topology implementation. This section presents various topology performance assessment and metrics.

5.1 Topology parameters

Various factors such as bisection width, diameter, degree, and link complexity are some of the parameters that distinguish and characterize one topology from the others. Some of these parameters are discussed in the following subsections.

5.1.1 Node degree

As aforementioned, a network can be regular or irregular if the entire node exhibits the same degree or not. The node degree is the number of edges connected to the node. Moreover, the node degree defines the node's I/O complexity, and depending on the topology, it can vary or constant with the network size. Also, topological features such as constant node degree and smaller degree are normally desirable for a more scalable network. For instance, the required effort in adding new nodes to the existing network is eased by the former feature while the latter facilitates less hardware cost on links. Besides, a constraint always exists on the node degree regarding the number of a node's direct neighbors. Furthermore, it is noteworthy that the node constraint increases due to the communication protocol and hardware limitations. These relate to node degree and port numbers that a node can support. Other factors such as network scalability and space complexity are performance considerations that limit effective node communication.

5.1.2 Diameter

In network topology, the diameter is the maximum shortest distance (path) between the node pairs. Also, in a situation where no direct connection exists between two nodes, the message from the source has to transfers through a number of intermediate nodes to get to its destination. Based on this, multiple hops delay is introduced. Moreover, this delay corresponds to the total number of hops to the destination. Consequently, in network topology, the maximum shortest path length is an important metric. In general, apart from its capability of providing predictable traffic flow and routing paths, a small diameter helps in offering low latency and facilitates network troubleshooting.

5.1.3 Link complexity

In a topology, link complexity defines the aggregate number of links or interconnects. It should be noted that the link complexity is proportional to the network scale and the highest complexity is presented by fully connected networks. Furthermore, when extra links are added to certain networks, their diameters reduce. This can help in offering better communication with lower latency between nodes. However, apart from the introduced complexity, additional links are expensive. Besides, high overhead (i.e. cost, area, etc.) and hardware complexity can also be induced by a high link complexity.

5.1.4 Bisection width

The bisection width is the minimum number of required edges that should be removed so as to divide a network topology into two halves (sub-network) with virtually equal size. It should be noted that a large bisection width is usually desirable for better network stability. This is due to the offered more paths between two sub-network entities and consequently helps in enhancing the overall performance. Also, a large bisection bandwidth can be achieved with a large bisection width. Bisection bandwidth, B_b, can be expressed as

$$B_b = W_b \times B_c \tag{1}$$

where W_b denotes the bisection width and B_c represents the communication channel's bandwidth.

5.2 Performance metrics

There are a number of metrics/parameters that can be employed for assessing the NoC's performance. Some of the performance metrics are presented in this section.

5.2.1 Max end-to-end latency

Latency is the time taken for delivering the packet from the source to the destination. Also, the maximum latency experienced by a given pair of source-destination nodes located at the farthest distance in the network is known as the *Max End-to-End latency* [4, 21], while the metric used to describe the lower latency bound, when there is no other network traffic is the *zero-load latency* [8]. Consider the wormhole switching based NoC, the zero-load latency can be expressed as [26].

$$T_z = \underbrace{\Gamma \cdot t_r}_{\text{Routing delay}} + t_c + \underbrace{L_p/B_c}_{\text{Serialization delay}} \tag{2}$$

where $t_r = t_a + t_s$ represents the router delay, with t_a being the arbitration logic delay and t_s, the switch delay, Γ denotes the average number of routers traversed by a packet to the destination node, t_c, denotes the propagation delay due to communication channel, and L_p represents the length of the packet in bits.

The average latency (delay) can be determined by taking the end-to-end delay mean of each successfully transmitted packet. In the computation of NoC performance, the average network latency is normally employed and can be expressed as [27].

$$T_{av} = \frac{\sum_{i=1}^{P} L_i}{P} \tag{3}$$

where P denotes the number of transmitted flits[8], L_i represents the network latency of the flit i.

It is noteworthy that in the estimation of average end-to-end latency, the packets lost during transmission are not taken into consideration. Also, how swift the packets can be delivered to their destinations indicates the average end-to-end latency, and the larger the value, the less efficient the network [21].

5.2.2 Dropping probability

When packets traverse in network topology, there may be packet loss due to network overloading and transmission errors. The packet loss can be determined by estimating the difference between the total sent packets by the source nodes and those received by the destination nodes. Similarly, the ratio of the dropped (lost) packets to the total sent packets by the source nodes is the dropping probability of the topology. The packet loss, P_l, and the dropping probability, D_p, can be expressed, respectively as [21].

$$P_l = \sum P_g - \sum P_r, \tag{4}$$

$$D_p = \frac{\sum P_d}{\sum P_g}, \tag{5}$$

where P_g denotes the generated packets by the source, P_r represents the packets received by the destination, and P_d is the dropped packets.

---

[8] The flits are fundamental packets for the execution of link flow control operations and synchronization between routers.

Moreover, owing to the QoS, a low dropping probability rate is preferred. For instance, 0 dropping probability implies no packet drop in the topology, while a 100 dropping probability denotes that the entire packets are dropped [4]. Also, there exist maximum acceptable loss rates for different applications.

5.2.3 Throughput

Throughput, χ, is the rate of packets that are delivered successfully to the destination nodes [4] and can be expressed as [27].

$$\chi = \frac{\sum_{i=1}^{S} P_i}{\tau\zeta} \tag{6}$$

where ζ represents the number of routing nodes, τ denotes the total execution time (total time taken), P_i represents the number of flits that information i contains in time τ, and s is the number of information sent or received in time τ.

Furthermore, the average throughput can be estimated by averaging the number of successfully received packets per second during transmission [21]. Also, when the traffic rate in the NoC is high, the traversing packets in the network will be contending, leading to transmission latency. Then, there will be an injection rate at which the latency will be prohibitively high. This instant is known as the *saturation throughput point* [8].

It is noteworthy that the throughput is mainly contingent on several parameters such as the flow control, employed routing algorithm, available signal-to-noise ratio, available bandwidth, data loss, hardware utilization, buffering, and employed protocol [8]. Also, throughput is based on link utilization in the network. This parameter signifies the number of supported flits by each link in unit time and can be defined as [27].

$$U_\ell = \sum_{i=1}^{s} P_i \tau \frac{\Gamma_{min}}{\Psi} \tag{7}$$

where Γ_{min} is the minimum number of routers traversed by data i and Ψ represents the number of links.

5.2.4 Average queue occupancy

The mean queue length measured as regards packets is the average queue length. Moreover, it can be used to indicates buffer utilization. Therefore, a shorter queue signifies a lower buffer utilization as well as shorter queuing delay. Also, to get the utilization ratio, the queue length is sampled at every time slot [21]. Furthermore, different active queue management techniques such as random early detection, deficit round-robin, fair queuing, drop-tail, stochastic fair queue, and random exponential marking can be employed for packet flow control between various source nodes and destination nodes. This can be achieved through the management of the intermediate routers' buffers [28].

6. NoC challenges

As aforementioned, NoC offers a scalable and modular platform for offering efficient on-chip communication for addressing the trend of SoC integration, however, certain related challenges still demand attention to further enhance the system performance. In this section, we discuss a number of on-chip challenges.

6.1 Links

The choice of parallel or serial links for the data transfer has been one of the primary issues in the NoC due to their associated features. For instance, serial links can considerably save the area, alleviate noise, and reduce interference. However, serializer and de-serializer circuits are required for data transport. On the other hand, the parallel link helps in reducing the power dissipation nevertheless, it consumes more area owing to its buffer-based architecture [10, 29].

6.2 Router architecture

One of the main factors for the embedded systems is the product cost. Besides, the underlying architecture is essential to be small in size and consequently consume less power. Based on this, the routing protocol design presents a tradeoff between cost and performance. For instance, router design will be complicated by a complex routing protocol. In this context, more area and power will be consumed, making it uneconomical. On the other hand, a simpler routing protocol will be a cost-effective solution however, its performance in traffic routing will not be effective [10, 29].

6.3 NoC area/space optimization

In the NoC architecture, communication takes place through the connected modules through the router network by means of long links. Besides, the various schemes such as link sizing, packet sizing, buffer sizing, flow/congestion control, and switching protocol for different topologies not only demand enormous space for NoC design but also make open benchmarks challenging. Consequently, to enhance system performance, link optimization is very imperative. Although the issue can be addressed with repeater implementation, more chip area will be consumed [10, 29]. Similarly, efficient design tools for space evaluation and implementation that can be seamlessly integrated with the current standard tools are required to facilitate extensive employment of NoC technology. Also, because of the complexity of NoC systems, network simulation time will be prohibitive. Therefore, to optimize the simulation speed, innovative techniques are required. Besides, to ensure appropriate architecture selection for an application, open benchmarks are required for different performance features comparison [12].

6.4 Latency

In NoCs, latency increase happens due to additional delay for data packetization/de-packetization at the NIs. It can also be attributed to the fault tolerance protocol overheads and flow/congestion control delays. Besides, owing to contention and buffering, routing delays can also affect network performance. Consequently, to enhance network performance (i.e. to satisfy the stringent latency constraints), native NoC support, low diameter topologies, and advanced flow control approaches are demanded [12].

6.5 Power consumption leakage

Depending on the application, the link utilization in the NoC may vary and in several cases, it is very low. To meet the worst-case scenario requirements, the NoCs are designed to keep redundant links and function at low link utilization. Nevertheless, even with ideal links, NoC consumes relatively much power owing to the associated complex routing logic blocks and NI. As a result, to further enhance its performance regarding leakage power consumption reduction, innovative architecture and circuit techniques are required [10, 29].

7. Simulation analysis and results

In this section, we consider a 4×4 2D mesh, torus, and fat-tree-based NoCs in the simulation analysis and present results regarding their performance. The simulation is based on the OPNET network simulator. We assumed that there is independent packet generation by the functional cores at time intervals that follow a negative exponential distribution. Also, we assumed a uniform traffic pattern where each processor forwards packets to others with equal probability. Likewise, we use payload packet sizes range from 256 to 1024 bytes for a diverse offered load. End-to-end (ETE) delay (latency) and throughput are the considered performance metrics.

There are general drift patterns in the considered NoCs, resulting in a similar performance exhibition. **Figure 12(a)** depicts the average latency and indicates it increases with an increase in the offered load and rises faster after saturation. For instance, with mesh topology, the average latency is less than 2 μs before the offered load of 0.2 then grows intensely later. This rapid increase after saturation load can be attributed to network congestion. Also, at 70 μs, the offered load for mesh, torus, and fat-tree are about 0.45, 0.48, and 0.56, respectively. Also, **Figure 12(a)** illustrates the throughput with various offered loads and shows it increases with an increase in the offered loads before saturation. For instance, at 0.6 offered load, the throughput of mesh, torus, and fat-tree are about 150, 160, and 180 Gbit/s, respectively.

Furthermore, the average latency considering various offered loads and packet lengths is illustrated in **Figure 13(a)**. It is noteworthy that based on the packet sizes, the network saturates at different loads. For instance, the average latency is comparatively low before the saturation load and a considerable surge occurs after it. So, the saturation load for 256 bytes is lower compared with larger packets. Also, with an increase in the packet size, the variation between the average latency curves of

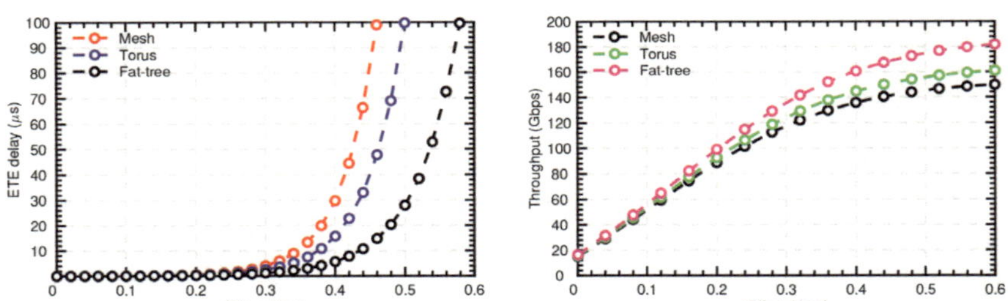

(a) average latency (b) throughput

Figure 12.
Performance analysis of the considered topologies under different offered loads and at 256 bytes packet.

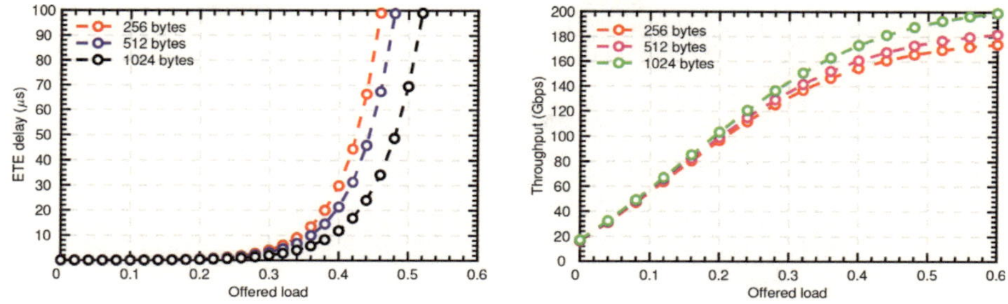

(a) average latency (b) throughput

Figure 13.
Performance analysis of 4×4 mesh network under different offered loads and packet lengths.

the adjacent packet sizes turns out to be smaller. Similarly, the network throughput based on different offered loads and packet sizes is illustrated in **Figure 13(b)**.

In general, larger packet length results in lower average latency and higher saturation load. This is owing to the required more transmission time and inter-packet arrival interval by the larger packets compared with smaller ones given the same offered load. Based on this, the path-setup packets blocking possibility can be minimized. Consequently, the destination can effectively receive more packets, resulting in higher throughput and saturation loads.

8. Conclusion

The current and the next-generation applications demand reliable and high-performance on-chip communication for various domain-specific/architecture-aware large-scale multiprocessor system-on-chips/embedded systems. Some of the major research areas in the NoC are topology, routing, and switching. In this chapter, we have presented a comprehensive overview of the evolution of its architectures and have highlighted their associated features. In this context, we have focused mainly on some defining features such as the topology, routing algorithms, and switching arrangements that are promising for the current and future on-chip architectures. Besides, we have presented different application mapping algorithms that are capable of influencing the NoC overall performance, mainly regarding the power requirement and network latency. Also, we have discussed various open problems in its design and implementation. It is noteworthy that the choice of NoC depends mainly on the use cases that will determine the trade-offs between the area, cost, power consumption, and overall performance.

Acknowledgements

This work is supported by the European Regional Development Fund (FEDER), and Internationalization Operational Programme (COMPETE 2020) of the Portugal 2020 (P2020) framework, under the projects DSPMetroNet (POCI-01-0145-FEDER-029405) and UIDB/50008/2020-UIDP/50008/2020 (DigCORE), and by FCT/MCTES through national funds and when applicable co-funded EU funds under the project UIDB/50008/2020-UIDP/50008/2020 (action QuRUNNER).

Author details

Isiaka A. Alimi[1*], Romil K. Patel[1,2], Oluyomi Aboderin[1], Abdelgader M. Abdalla[1], Ramoni A. Gbadamosi[3], Nelson J. Muga[1], Armando N. Pinto[1,2] and António L. Teixeira[1,2]

1 Instituto de Telecomunicações and University of Aveiro, Portugal

2 Department of Electronics, Telecommunications and Informatics, University of Aveiro, Portugal

3 Faculty of Science, National Open University of Nigeria, Akure, Nigeria

*Address all correspondence to: iaalimi@ua.pt

References

[1] A. Kalita, K. Ray, A. Biswas, and M. A. Hussain. A topology for network-on-chip. In *2016 International Conference on Information Communication and Embedded Systems (ICICES)*, pages 1–7, 2016.

[2] Haytham Elmiligi, Ahmed A. Morgan, M. Watheq El-Kharashi, and Fayez Gebali. Power optimization for application-specific networks-on-chips: A topology-based approach. *Microprocessors and Microsystems*, 33(5):343–355, 2009.

[3] D. Bertozzi and L. Benini. Xpipes: a network-on-chip architecture for gigascale systems-on-chip. *IEEE Circuits and Systems Magazine*, 4(2):18–31, 2004.

[4] T. N. Kamal Reddy, A. K. Swain, J. K. Singh, and K. K. Mahapatra. Performance assessment of different Network-on-Chip topologies. In *2014 2nd International Conference on Devices, Circuits and Systems (ICDCS)*, pages 1–5, 2014.

[5] Zulqar Nain, Rashid Ali, Sheraz Anjum, M. Afzal, and S. Kim. A Network Adaptive Fault-Tolerant Routing Algorithm for Demanding Latency and Throughput Applications of Network-on-a-Chip Designs. *Electronics*, 9(1076):1–18, 2020.

[6] Isiaka A. Alimi, Ana Tavares, Cátia Pinho, Abdelgader M. Abdalla, Paulo P. Monteiro, and António L. Teixeira. *Enabling Optical Wired and Wireless Technologies for 5G and Beyond Networks*, chapter 8, pages 1–31. IntechOpen, London, 2019.

[7] Cátia Pinho, Isiaka Alimi, Mário Lima, Paulo Monteiro, and António Teixeira. *Spatial Light Modulation as a Flexible Platform for Optical Systems*, chapter 7, pages 1–21. IntechOpen, London, 2019.

[8] Haseeb Bokhari and Sri Parameswaran. *Network-on-Chip Design*, chapter 5, pages 461–489. Springer Netherlands, Dordrecht, 2017.

[9] G. Passas, M. Katevenis, and D. Pnevmatikatos. Crossbar NoCs Are Scalable Beyond 100 Nodes. *IEEE Transactions on Computer-Aided Design of Integrated Circuits and Systems*, 31(4): 573–585, 2012.

[10] H. J. Mahanta, A. Biswas, and M. A. Hussain. Networks on Chip: The New Trend of On-Chip Interconnection. In *2014 Fourth International Conference on Communication Systems and Network Technologies*, pages 1050–1053, 2014.

[11] S. Johari and V. Sehgal. Master-based routing algorithm and communication-based cluster topology for 2D NoC. *The Journal of Supercomputing*, 71:4260–4286, 2015.

[12] Sudeep Pasricha and Nikil Dutt. *Networks-On-Chip*, chapter 12, pages 439–471. Systems on Silicon. Morgan Kaufmann, Burlington, 2008.

[13] B. N. K. Reddy, D. Kishan, and B. V. Vani. Performance constrained multi-application network on chip core mapping. *International Journal of Speech Technology*, 22:927–936, 2019.

[14] Evgeny Bolotin and Israel Cidon and Ran Ginosar and Avinoam Kolodny. Cost considerations in network on chip. *Integration*, 38(1):19–42, 2004.

[15] D. Flynn. AMBA: enabling reusable on-chip designs. *IEEE Micro*, 17(4): 20–27, 1997.

[16] Rohita P. Patil and Pratima V. Sangamkar. Review of System-On-Chip Bus Protocols. *International Journal of Advanced Research in Electrical, Electronics and Instrumentation Engineering*, 4(1):271–281, 2015.

[17] María Dolores Valdés Peña Juan José Rodríguez Andina, Eduardo de la Torre Arnanz. *Embedded Processors in FPGA Architectures*.

[18] Mohandeep Sharma and D. Kumar. Wishbone bus architecture - a survey and comparison. *International Journal of VLSI design & Communication Systems (VLSICS)*, 3(2):107–124, 2012.

[19] Altera. Avalon Bus Specification. Reference manual: 2.3, July 2003.

[20] Tobias Bjerregaard and Shankar Mahadevan. A Survey of Research and Practices of Network-on-Chip. *ACM Comput. Surv.*, 38(1):1–es, June 2006.

[21] J. Chen, P. Gillard, and Cheng Li. Network-on-Chip (NoC) Topologies and Performance: A Review. In *Proceedings of the 2011 Newfoundland Electrical and Computer Engineering Conference (NECEC)*, pages 1–6, 2011.

[22] Y. Li and S. Peng. Dual-cubes: A new interconnection network for high-performance computer clusters. In *International Computer Symposium, Workshop on Computer Architecture*, pages 1–7, 2000.

[23] S. Ma, N. E. Jerger, and Z. Wang. DBAR: An efficient routing algorithm to support multiple concurrent applications in networks-on-chip. In *2011 38th Annual International Symposium on Computer Architecture (ISCA)*, pages 413–424, 2011.

[24] W. Amin, F. Hussain, S. Anjum, S. Khan, N. K. Baloch, Z. Nain, and S. W. Kim. Performance Evaluation of Application Mapping Approaches for Network-on-Chip Designs. *IEEE Access*, 8:63607–63631, 2020.

[25] Pradip Kumar Sahu and Santanu Chattopadhyay. A survey on application mapping strategies for Network-on-Chip design. *Journal of Systems Architecture*, 59(1):60–76, 2013.

[26] V. F. Pavlidis and E. G. Friedman. 3-D Topologies for Networks-on-Chip. *IEEE Transactions on Very Large Scale Integration (VLSI) Systems*, 15(10): 1081–1090, 2007.

[27] Xingang Ju and Liang Yang. Performance analysis and comparison of 2×4 network on chip topology. *Microprocessors and Microsystems*, 36(6): 505–509, 2012.

[28] S. Patel and A. Sharma. The low-rate denial of service attack based comparative study of active queue management scheme. In *2017 Tenth International Conference on Contemporary Computing (IC3)*, pages 1–3, 2017.

[29] A. Agarwal and R. Shankar. Survey of Network on Chip (NoC) Architectures & Contributions. volume 3, pages 21–27, 2009.

Section 3

Microstructure Fabrication and Routing Optimization

Chapter 4

A Novel Approach for the Design of Fault-Tolerant Routing Algorithms in NoCs: Passage of Faulty Nodes, Not Always Detour

Masaru Fukushi and Yota Kurokawa

Abstract

Due to the faults in system fabrication and run time, designing an efficient fault-tolerant routing algorithm with the property of deadlock-freeness is crucial for realizing dependable Network-on-Chip (NoC) systems with high communication performance. In this chapter, we introduce a novel approach for the design of fault-tolerant routing algorithms in NoCs. The common idea of the fault-tolerant routing has been undoubtedly to detour faulty nodes, while our approach allows passing through faulty nodes with the slight modification of NoC architecture. As a design example, we present an XY-based routing algorithm with the passage function. To investigate the effect of the approach, we compare the communication performance (i.e. average latency) of the XY-based algorithm with well-known region-based algorithms under the condition of with and without virtual channels. Finally, we provide possible directions of future research on the fault-tolerant routing with the passage function.

Keywords: network-on-chip (NoC), fault-tolerant routing, two-dimensional mesh, passage, dependability

1. Introduction

Demand for computation power will never stop, and it is ever increasing year by year in a variety of scientific research fields. As can be seen in the modern multi-processor system-on-chips and many core systems [1–3], this makes computing hardware devices equip with hundreds or thousands of processor cores for providing high computation power by parallel processing on a chip. For the implementation of such highly-integrated parallel systems, Network-on-Chip (NoC) has emerged as a promising paradigm. In NoCs, each node (i.e. a processor core with a router) is connected by an on-chip network and communication among them are done by transferring packets on the network. Using global interconnection structure reduces the difficulty of wiring design and latency of signal transmission and offers high scalability, in comparison with point-to-point signal wires or shared busses [4].

One of the most important and fundamental issues that must be addressed for NoCs is fault-tolerant routing. Definitely, routing of packets plays a key role in parallel systems because it has significant impact on the overall system performance. Meanwhile, the occurrence of faults during system fabrication and run time is inevitable, and it is almost impossible to completely remove their adverse effects from the systems even if some redundancy is incorporated. A single faulty node disrupts packet routing between many pairs of nodes, resulting in the failure of the entire system. Besides, a deadlock (i.e. circular waiting of packets) will occur if an adopted routing algorithm is imperfect. Once the deadlock occurs, packets can never proceed to the destinations, and thus resulting in the malfunction of the entire system. Therefore, designing an efficient fault-tolerant routing algorithm with the property of deadlock-freeness is crucial for realizing dependable NoC systems with high communication performance.

So far, extensive research has been devoted to fault-tolerant routing not only for NoCs but for traditional parallel computers. Although there exist several basic approaches, as we reviewed in Section 2, the common idea of the fault-tolerant routing remains unchanged from the earliest, and it has been undoubtedly to detour faulty nodes. This is quite natural because the purpose of the fault-tolerant routing is to route packets from source to destination nodes without entering faulty parts. Meanwhile, it is also obvious that detouring faulty nodes increases the communication latency as the packet is misrouted apart from the minimal path to the destination. One may consider that the increase in the communication latency is very little. This is true if packets are routed without interfered by other packets. However, it can be substantial increase under the situation where a number of packets are routed simultaneously and thus frequently blocked by others.

In this chapter, we introduce a novel approach for the design of fault-tolerant routing algorithms. In contrast to the common idea of detouring faulty nodes, our approach allows passing through them with the slight modification of NoC architecture. We provide a general methodology for designing a fault-tolerant routing algorithm with the passage of faulty nodes. As a design example, we describe an XY-based routing algorithm with the passage function. By computer simulations, we reveal the communication performance of the algorithm under the condition of with and without Virtual Channels (VCs), in comparison with well-known region-based routing algorithms.

The rest of this chapter is organized as follows: Section 2 presents the architecture of NoC, the basis of packet routing, and the related works of fault-tolerant routing algorithms. Section 3 presents the basic idea of the proposed approach and XY-based fault-tolerant routing algorithm, inclusive of the proof of the deadlock-freeness. Section 4 presents the results of the performance evaluation. Finally, Section 5 concludes the chapter with some possible direction of future research.

2. NoC architecture and fault-tolerant routing

2.1 2D mesh NoC

Target NoC topology in this chapter is a popular 2D mesh which has nodes of m rows and n columns. **Figure 1** shows the general architecture of the 2D mesh NoC. Each node is composed of a processor core and a router. The processor core runs assigned computation tasks, which can be either independent one or a part of parallel programs, while the router forwards packets to one of the neighbor routers or its local processor core to support the communication among cores. Each node has a unique address (i,j), where $i \in X = \{1, 2, \cdots, m\}$ and $j \in Y = \{1, 2, \cdots, n\}$. In the 2D mesh NoC, a node (i,j) is connected to at most four neighbor nodes, $(i \pm 1, j)$ and $(i, j \pm 1)$, via two unidirectional links, if $i \pm 1 \in X$ and $j \pm 1 \in Y$. For the ease of

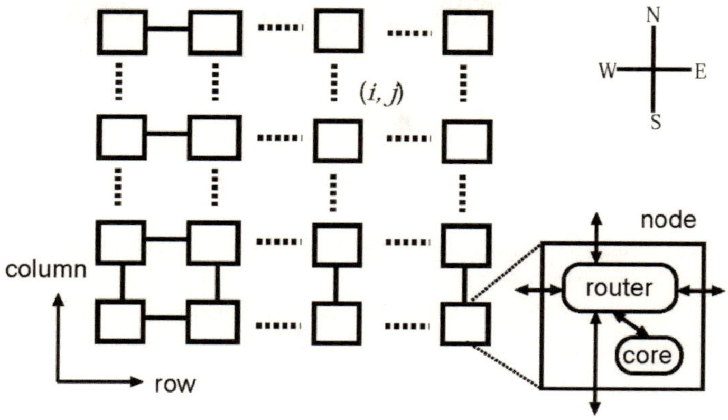

Figure 1.
Architecture of 2D mesh NoC.

explanation, positive/negative directions of x (row) and y (column) axes are called east/west and north/south, respectively.

Figure 2 shows the block diagram of the router. In the typical wormhole routing adopted in NoCs, a packet is divided into a sequence of fixed-size units of data, called flits, and transferred by routers one after another. Each router consists of five input/output units, five routing circuits, a VCs allocator, a crossbar switch, and a switch allocator. When a head flit (i.e. a flit having routing information) is transferred to a router and stored into a buffer in the input unit, the following processes are applied.

1. An output port to which the flit is forwarded is determined by the routing circuit.

2. A VC (i.e. buffer) to be used is determined by the VC allocator.

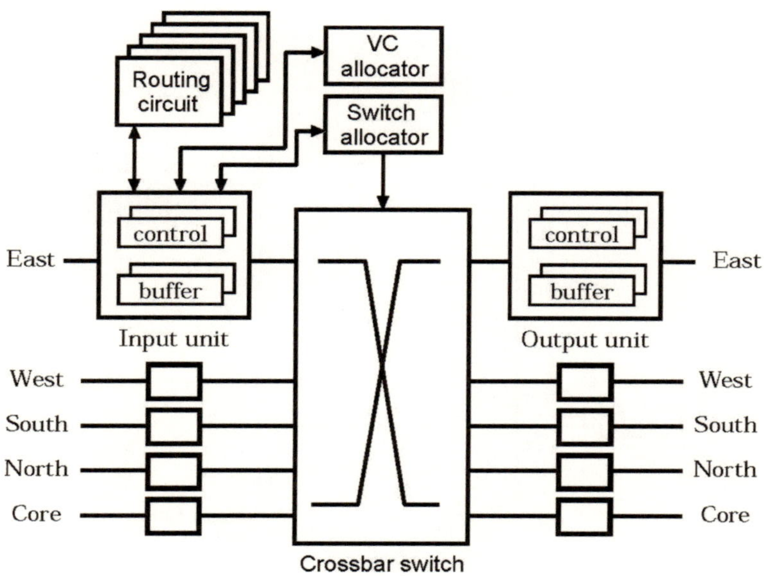

Figure 2.
Architecture of router.

3. The crossbar switch is set up by the switch allocator to connect the input unit and the output unit associated with the determined output port.

4. The flit is moved from the input to the output units.

5. Finally, the flit is forwarded to the corresponding input unit of the next router.

The incoming head flit moves to the next router at the fifth cycle if there are no contentions, and the subsequent flits follow it in a pipeline fashion. This is a standard five-cycle router [5]. If no VC is used in an adopted routing algorithm, the router is reduced to a four-cycle router, as the second process (i.e. VC allocation) is omitted.

2.2 Deadlock

In routing packets in accordance with a routing algorithm, the algorithm must care about the occurrence of deadlocks. Deadlock is a situation where packets wait on one another to release the buffers. **Figure 3** shows an example of a deadlock. In this example, a packet A is routed to the node (2, 1) via (1, 1), which is blocked by a packet B at (1, 1). The packet B is also routed to (2, 2) via (2, 1), which is blocked by a packet C. The packets C and D are also routed similarly, but blocked by the packets D and A, respectively, resulting in circular waiting of packets. Once a deadlock occurs, packets involved in the circular waiting cannot proceed toward the destinations forever. Therefore, deadlock-freeness must be guaranteed in the routing algorithm.

There have been two approaches to preventing deadlocks; approaches with and without VCs. In the approach with VCs, the original network is multiplexed into several virtual networks by VCs. For example, in **Figure 3**, if packets A and C are supposed to be routed on a virtual network with a VC and packets B and D are on a different virtual network with other VC, then the circular waiting is decomposed and packets get to proceed to the destinations. In the approach without VCs, a routing algorithm is carefully designed so that deadlocks never occur in the original

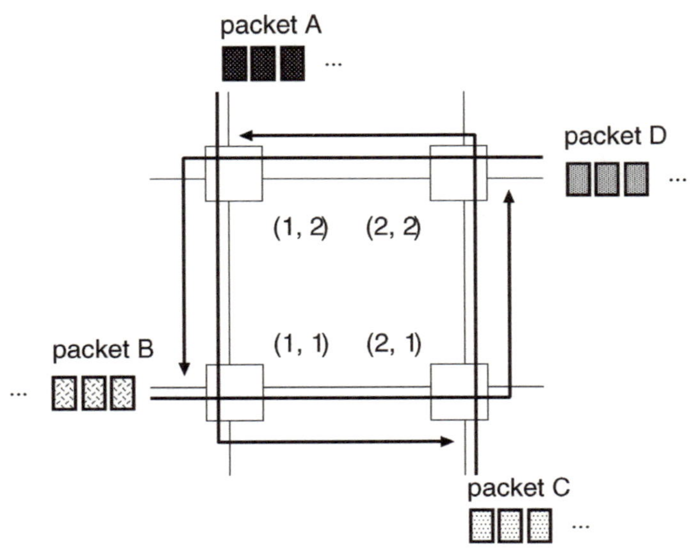

Figure 3.
Example of a deadlock.

physical network. For example, in **Figure 3**, if packets A and C are routed via (2, 2) and (1, 1), respectively, i.e., forced to move in x-direction first, then no deadlocks occur. This approach has an area advantage over the former approach because the implementation of VCs involves the replication of buffers and control circuits in all input/output units in all routers.

2.3 Related works

Fault-tolerant routing has been the subject of extensive research not only for NoCs but for traditional parallel computers over the past few decades. Most of the existing fault-tolerant routing algorithms for 2D mesh networks fall into the following three categories: (1) those employ a routing table, (2) those relax the constraints of guaranteed delivery or deadlock-freeness, and (3) those define some form of fault information on routers and detour paths.

In the first category, a routing table is employed in each router to route packets to the destinations. Routing tables contain routing information such as next hops for destinations, status of the network, and/or fault information. Hsin et al. [6] proposed an algorithm which employs ant colony optimization for traffic balancing. Liu et al. [7] proposed an algorithm which introduces coarse and fine-grained look-ahead schemes to obtain the information of other routers within the range of four hops. This algorithm requires two VCs for each input/output port to route packets. Zhao et al. [8] proposed an algorithm to provide minimal paths using the information of whole network. In general, those algorithms offer flexible route selection; however, they require a large amount of circuits to implement a routing table and complex calculation mechanism to create/update the table in all routers.

In the second category, constraints of guaranteed delivery or deadlock-freeness is relaxed to ease the design of routing algorithms. Janfaza et al. [9] proposed an adaptive routing algorithm which employs timeout and packet reinjection. Information of intermediate nodes is recorded in each packet and two VCs are used to route packets. Sinha et al. [10] proposed an algorithm based on the common XY and YX routing. This algorithm allows U-turn using several VCs. Wang et al. [11] proposed an algorithm which relaxes transmission accuracy for the applications that allow lossy communication. This algorithm discards conflicting approximate flits without retransmission and recovers them after packet transmission. Those algorithms are imperfect in that 100% packet reachability or deadlock-freeness are not guaranteed by the routing algorithms. Retransmission of packets generally results in a high communication latency.

In the third category, some form of fault information is defined for routers to detour faulty parts. Usually, clusters of faulty nodes, called fault blocks, are defined in the networks with the detour paths. Chen et al. [12], Holsmark et al. [13], Fu et al. [14], and Fukushima et al. [15] proposed routing algorithms which generate rectangular fault blocks and detour them without using VCs. Wu [16] and Chalasani et al. [17] proposed routing algorithms which can deal with convex and nonconvex fault blocks, respectively. In [17], four VCs are used to choose shorter detour paths. Those algorithms called region-based algorithms provide simple but strict routing rules to guarantee the deadlock-freeness and 100% packet reachability, and thus, they can be implemented as a small circuit in the routing circuit of each router. They are practical and suitable for NoCs. However, one drawback is that fault blocks may include several non-faulty nodes, which are to be deactivated (i.e. unused nodes); therefore, the number of unused nodes and the length of detour paths are prone to increase if there exist a number of faulty nodes in the network.

Although extensive research has been devoted to designing fault-tolerant routing algorithms inclusive of the above ones, the common idea remains unchanged from the earliest, and it has been undoubtedly to detour faulty nodes. If a packet must detour a faulty node (i,j), the hop count between $(i-1,j)$ and $(i+1,j)$ is increased by two, which increases the communication latency by at least ten cycles in an NoC with five-cycle routers. This can be substantial increase in the situation where the network gets congested (i.e. by packet blocking) or includes a number of faulty nodes (i.e. by detouring). This is a serious problem for a large-scale parallel system on a single VLSI chip.

3. Proposed method

3.1 Basic approach and NoC architecture

Motivated by the problem presented in the previous section, we introduce a novel approach based on the opposite idea of the common approach; our approach allows packets to pass through the faulty nodes with slight modification of NoC architecture (originally proposed in [18]). Basic idea behind this approach is to reduce communication latency by saving detouring as much as possible.

Figure 4 shows the modified NoC architecture for supporting the proposed approach. Four electrical switches, bypass links, buffers to store one flit are added around each router. Each switch has two states, either normal or passage, as shown in this figure. In the state of passage, packets from the neighbor node are input to the bypass link not to the node. The switch states can be determined easily, once the node is tested and judged as faulty or not. In other words, they are determined so that the node becomes passage state if it is faulty or remains normal state otherwise. It is worth to note that buffers can be removed if packets are transmitted between routers in an asynchronous way.

3.2 Design methodology for fault-tolerant routing algorithms

Here, we provide a design methodology for fault-tolerant routing algorithms based on the passage of faulty nodes.

First, we clarify the fault model. A common assumption is made for faults [6, 12–16, 18, 19]; that is, permanent faults are considered to be associated only with nodes. In practice, the probabilities of links, switches, and buffers being faulty are not zero, though they will be substantially small because of the simplicity of their

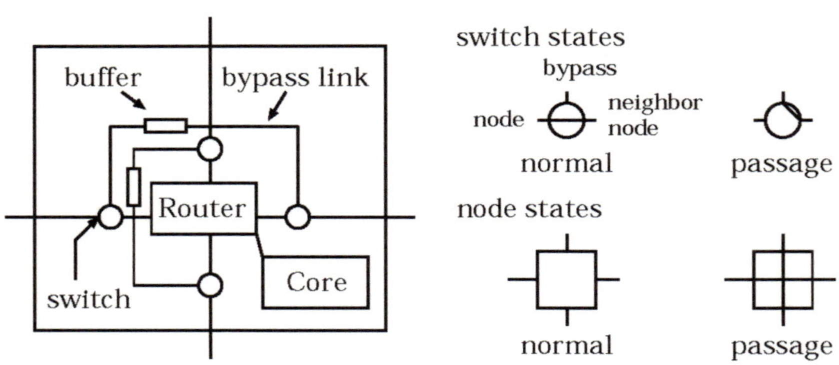

Figure 4.
Modified NoC architecture.

circuits [19]. For the faults on those circuits, one can employ some popular redundancy technique such as duplication and triplication if necessary.

Below is the general methodology for designing fault-tolerant routing algorithms with the passage function.

Step 1 Choose a base routing algorithm from the existing algorithms or design a new one. This algorithm is not necessary to be fault-tolerant, but should be deadlock-free.

Step 2 Decide which faulty nodes can be passed through and define routing rules for the remaining faulty nodes to be detoured. The resultant routing algorithm, denoted by R, is a candidate for the final algorithm.

Step 3 Verify if the candidate routing algorithm R is deadlock-free or not. If not, return to Step 2 to modify R.

Step 4 Repeat Steps 2 and 3 until a fault-tolerant and deadlock-free routing algorithm R is obtained.

3.3 Routing algorithm based on XY routing

As a design example, we introduce a new fault-tolerant routing algorithm based on the popular dimension order routing (i.e. XY routing for 2D meshes) [18]. In the following, we explain the details of each step in the design methodology.

In Step 1, we choose XY routing as a base routing algorithm. In XY routing, packets first proceed along x-direction until they reach the nodes having the same x-coordinates as the destinations, then proceed to the destinations along y-direction without changing the x-coordinates.

In Step 2, we must consider the case where passage must be restricted. For example, suppose that a packet moves from node $(i-1,j)$ to $(i+1,j)$ passing through a faulty node (i,j). If the destination node is (i,j') where $j' \neq j$, the packet keeps moving between $(i-1,j)$ and $(i+1,j)$ because the x-coordinate of the current node will never be the same as that of the destination node. The same kind of thing never happens in the y-direction. Therefore, we allow packets passing through faulty nodes only in the y-direction and let them detour faulty nodes through the south side in the x-directional movement. (This restriction is relaxed a bit in the final routing algorithm).

Then, we need to consider the case where a faulty node is on the south boundary of the network. In this case, packets cannot detour it through the south side, as they face the south boundary. To cope with this, we give the following definitions.

Definition 1 A faulty node (i,j) which is on the south boundary of a mesh network is defined as a South Faulty (SF) node, where $j = 0$.

Definition 2 A faulty node (i,j) which exists in the eight neighbor of any SF node (i',j') is also defined as an SF node, where $(i'-1 \leq i \leq i'+1)$ and $i \in X$, and $(j'-1 \leq j \leq j'+1)$ and $j \in Y$.

The process in Definition 2 is repeated until no SF nodes are generated. For SF nodes, we give a new routing rule such that packets must detour them through the north side.

In Step 3, we check the deadlock-freeness of the resultant routing algorithm R where packets detour faulty nodes/SF nodes through south/north side in the x-directional movement of XY routing and always pass through faulty nodes in the y-directional movement. Unfortunately, it is not hard to find the case where a deadlock occurs. **Figure 5** illustrates the example of a possible deadlock. Packets generated at nodes S1/S2 detour faulty and SF nodes in accordance with the routing algorithm R, but finally they are blocked by each other, resulting in circular waiting. Note that, generally, the deadlock in **Figure 5** can be occurred by more than two packets.

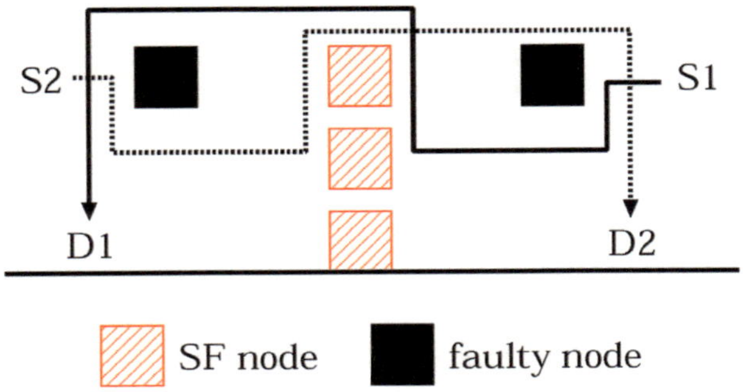

Figure 5.
Example of a possible deadlock.

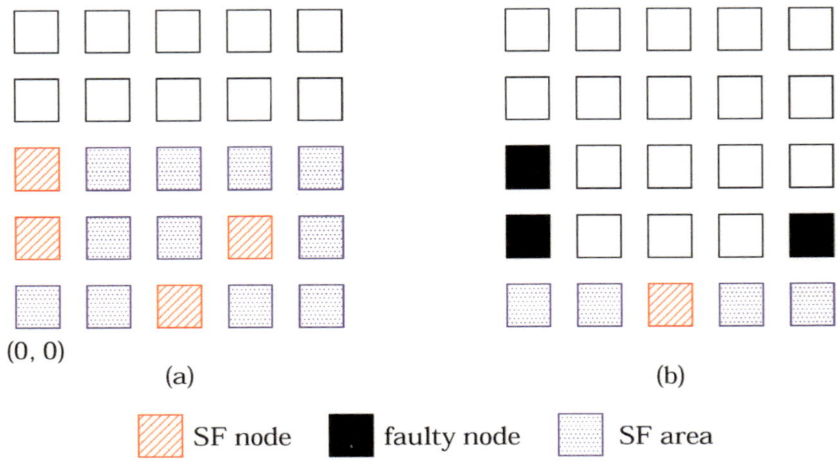

Figure 6.
SF area for xy-based routing algorithm.

To cope with the deadlock, we give the following definitions.

Definition 3 Let (i', j') be the coordinates of the north most SF node generated by repeating Definition 2. SF area is defined as the area consisting of all nodes (i,j) such that $j \leq j'$ for any $i \in X$. All faulty nodes in the SF area are changed to SF nodes.

For the newly generated SF nodes in Definition 3, the processes in Definitions 2 and 3 are repeated until no SF nodes are generated.

Figure 6 illustrates examples of the SF area. In the case of **Figure 6** (a), faulty node $(2, 0)$ is changed to an SF node by Definition 1 and subsequently $(3, 1)$ is changed to an SF node by Definition 2. Then, faulty node $(0, 1)$ on the west boundary is included in the SF area and thus changed to an SF node by Definition 3. Finally, faulty node $(0, 2)$ is changed to an SF node by Definition 2. According to the above processes, the SF area is configured as shown in the figure. In the case of **Figure 6** (b), faulty node $(4, 1)$ is not included in the SF area; hence, faulty nodes $(0, 1)$ and $(0, 2)$ are not changed to SF nodes.

By the above definitions, the deadlock in **Figure 5** can be solved. By Definition 3, two faulty nodes in **Figure 5** are changed to SF nodes and the SF area is defined as shown in **Figure 7**. Then, two packets detour the SF nodes, not faulty nodes, through the north side and get to proceed to the destinations as shown in the figure.

Figure 8 describes the finally obtained proposed routing algorithm. In this figure, C and D represent a current and a destination node, respectively. The

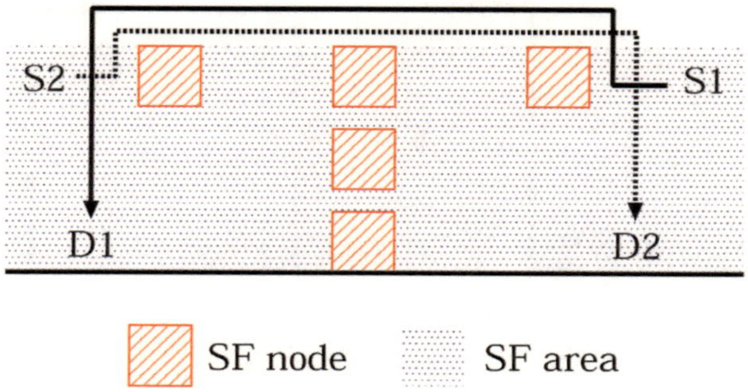

Figure 7.
Routing example without deadlocks.

proposed routing algorithm allows packets to pass through faulty and SF nodes in the movement of x-directions only if C and D are on the same row (i.e. lines 8 and 18 in **Figure 8**), while it always allows passage in the movement of y-directions.

Next, we prove the deadlock-freeness of the proposed algorithm described in **Figure 8**. First, we define turns of packets.

Definition 4 ES turn is a turn in which an incoming packet from the East neighbor is sent to the South neighbor at a router. Other seven turns are also defined similarly as shown in **Figure 9**.

Theorem 1 The routing algorithm in **Figure 8** is deadlock-free.

Proof. We prove that circular waiting of packets never occurs in both clockwise and counter-clockwise directions.

For the clockwise direction, we show that an SW turn is never aligned with an NE turn. The SW turn occurs in a non-SF area; however, the NE turn never occurs

```
1   XY-based fault-tolerant routing(C, D)
2   if (C is D)
3     consume the packet
4   elsif (D is to the west of C)
5     if (west neighbor is faulty)
6       if (west neighbor is an SF node)
7         Next_Route = North
8       elsif (C and D are on the same row)
9         Next_Route = West
10      else
11        Next_Route = South
12    else
13      Next_Route = West
14  elsif (D is to the east of C)
15    if (east neighbor is faulty)
16      if (east neighbor is an SF node)
17        Next_Route = North
18      elsif (C and D are on the same row)
19        Next_Route = East
20      else
21        Next_Route = South
22    else
23      Next_Route = East
24  elsif (C and D are on the same column)
25    if (D is to the north of C)
26      Next_Route = North
27    else
28      Next_Route = South
```

Figure 8.
A pseudo-code of the proposed routing algorithm.

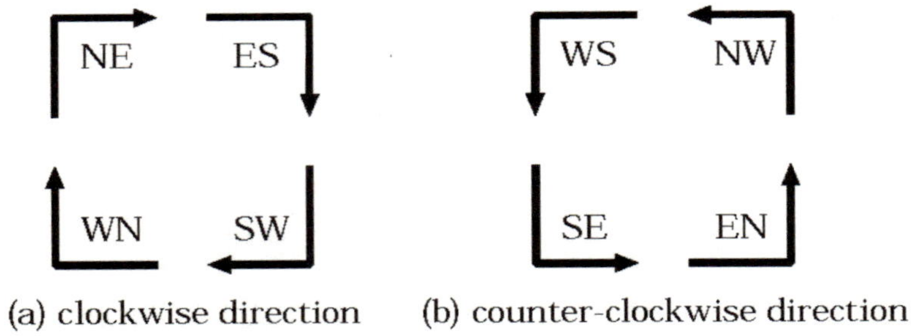

Figure 9.
Possible eight turns of packets.

in the area because it only occurs in an SF area. Conversely, the NE turn occurs in an SF area; however, the SW turn only occurs in a non-SF area. From the above, circular waiting never occurs in the clockwise direction.

For the counter-clockwise direction, we omit the proof because it is symmetrical to the proof for the clockwise direction.

Thus, the proposed routing algorithm is proved to be deadlock-free. □

4. Performance evaluation

4.1 Evaluation condition

To investigate the effect of the proposed approach, we have conducted computer simulations with a cycle-accurate custom simulator developed in C. This simulator accurately simulates the behavior of flits in all routers in a 2D mesh NoC. As explained in Section 2, if there are no contentions, each flit takes five (or four) cycles to move to the next node when VCs are used (or not used) in the adopted routing algorithm. Note that, as flits are transmitted in a pipeline fashion, a subsequent flit moves to the next node one cycle after the movement of the precedent flit if buffer space is available in the input unit of the router. It also takes one cycle to pass through a faulty node, as a buffer is placed on the bypass link.

Following three methods are evaluated in the simulations with the parameters listed in **Table 1**.

- Fukushima's method [15]: packets detour *rectangular* fault blocks with no additional VCs (denoted by M_r).

Parameter	Value	Unit
Network size	10×10	Nodes
Fault rate (f)	2, 4, 6, 8, 10	%
Packet length	16	Flits
Packet generation rate (p)	$0.05 \sim 1.0$	Packets/cycle/network
Input (Output) buffer depth	8 (1)	Flits
Simulation (Stabilization) cycle	50,000 (5000)	Cycles

Table 1.
Simulation parameters.

- Chalasani's method [17]: packets detour *nonconvex* fault blocks, such as L, T, and + shapes, as well as rectangular one using four VCs per link (denoted by M_{nc}).

- Our method [18]: packets can *pass through* or detour faulty nodes with no additional VCs (denoted by M_p).

The number of VCs required for each algorithm is different, and VCs can also be employed in the algorithms which require no VCs for the purpose of congestion avoidance. We use the notation of M-n to indicate the number of VCs (i.e. buffers), where M is either M_r, M_{nc} or M_p and n represents the number of VCs. For example, M_{nc}-4 denotes Chalasani's method with four VCs; M_p-1 denotes our proposed method with one buffer (i.e. no additional VCs).

In the simulations, faulty nodes are generated randomly according to the fault rate f, and packets are also generated randomly at each cycle according to the packet generation rate p during the simulation time of 50,000 cycles. Latency is not measured up to 5000 cycles to stabilize the network. The same fault patterns are used for all methods for fair comparison. The above trial is repeated 1000 times and the following metrics are measured.

Average latency is defined by the average cycles required for packets from the generation to the arrival.

Average node utilization rate is defined as the percentage of available nodes among all non-faulty nodes, i.e., given by $\{mn(1-f)-u\}/mn(1-f)$, where m and n is the number of rows and columns, respectively, f is the fault rate, and u is the number of unused nodes.

To make a quantitative evaluation of average latency, we define maximum latency reduction rate of an algorithm M_a for an algorithm M_b by

$$R(M_a, M_b) = max_p\, r_p(M_a, M_b), \tag{1}$$

where $r_p(M_a, M_b)$ represents latency reduction rate of M_a for M_b at the packet generation rate p and is defined by the following expression.

$$r_p(M_a, M_b) = \frac{L_b - L_a}{max\,(L_a, L_b)} \times 100, \tag{2}$$

where L_a and L_b is the average latency of M_a and M_b at p, respectively, and $max\,(L_a, L_b)$ is a function to return the larger of L_a and L_b.

4.2 Evaluation results

4.2.1 Overall trend

Figures 10–14 show the average latency as a function of packet generation rate p for each fault rate f. In the figures, x axis represents p, and a larger value indicates a higher request load; meanwhile, y axis represents average latency, and a larger value indicates higher delay in delivery of packets. When p is relatively low, the average latency of three algorithms is almost the same. On the other hand, when it is high, the difference becomes significant. The average latency of M_p and M_{nc} is smaller than that of M_{nc} and M_r, respectively, regardless of f and the number of VCs. M_p outperforms M_{nc} without using VCs, indicating that passage of faulty nodes has a significant impact on reducing average latency. As f increases, the

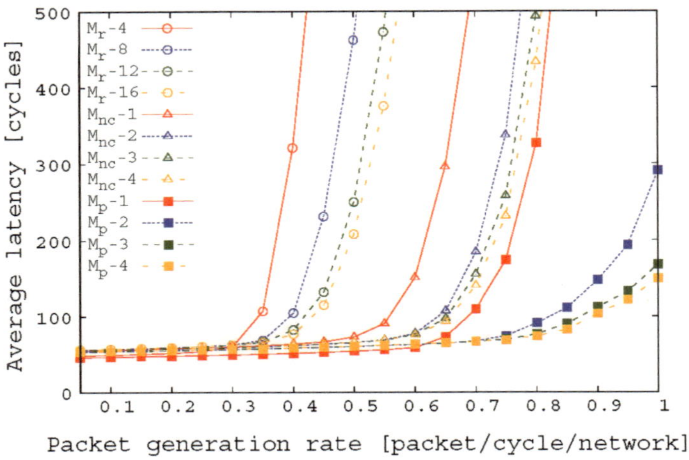

Figure 10.
Average latency for $f = 2\%$.

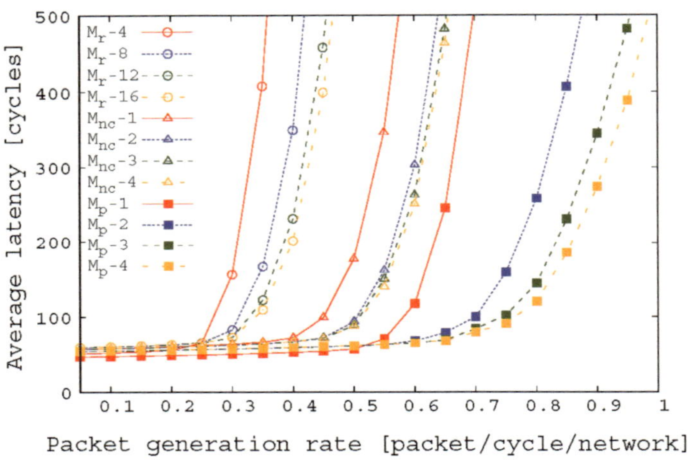

Figure 11.
Average latency for $f = 4\%$.

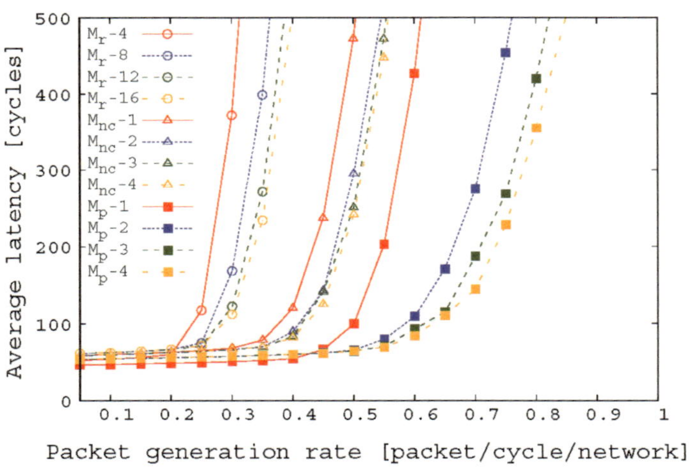

Figure 12.
Average latency for $f = 6\%$.

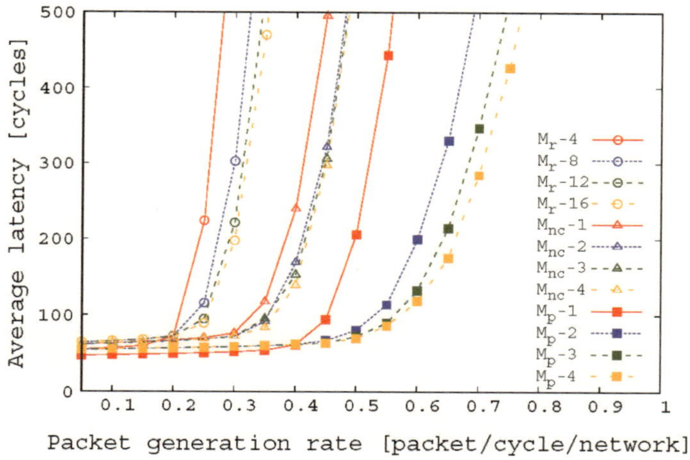

Figure 13.
Average latency for $f = 8\%$.

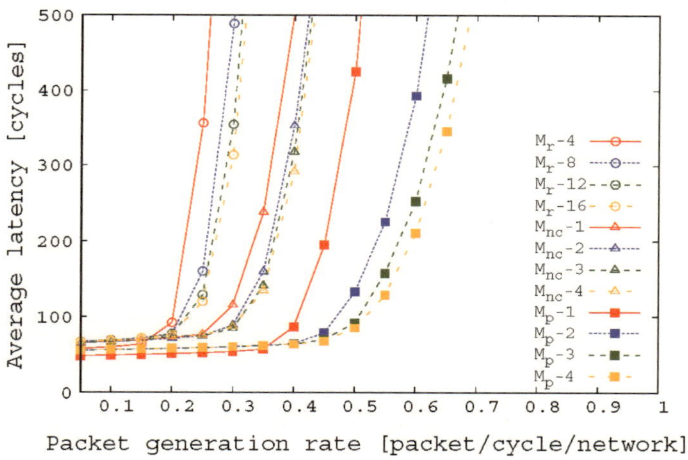

Figure 14.
Average latency for $f = 10\%$.

average latency of those algorithms is increased due to the increased number of faulty nodes.

Figure 15 shows the average node utilization rate. M_r and M_{nc} generate rectangular and nonconvex fault blocks, and accordingly, about 7% and 3% of non-faulty nodes become unused nodes, respectively. This is a cause of longer detour paths. Meanwhile, M_p does not generate any fault blocks and always keeps 100% utilization rate.

For the results shown in **Figures 10–14**, we make performance comparison of the routing algorithms in the following three conditions.

4.2.2 Performance comparison of the original routing algorithms

The average latency of the original routing algorithms is compared numerically (i.e. comparison of M_r-1, M_{nc}-4, and M_p-1). **Table 2** shows the maximum reduction rate of M_r-1 and M_p-1 for M_{nc}-4. The value of p at which the maximum reduction rate is attained is noted in parenthesis. As we saw in **Figures 10–14**, average latency of M_r is higher than that of M_{nc}; hence, all rates of $R(M_r\text{-}1, M_{nc}\text{-}4)$ are negative

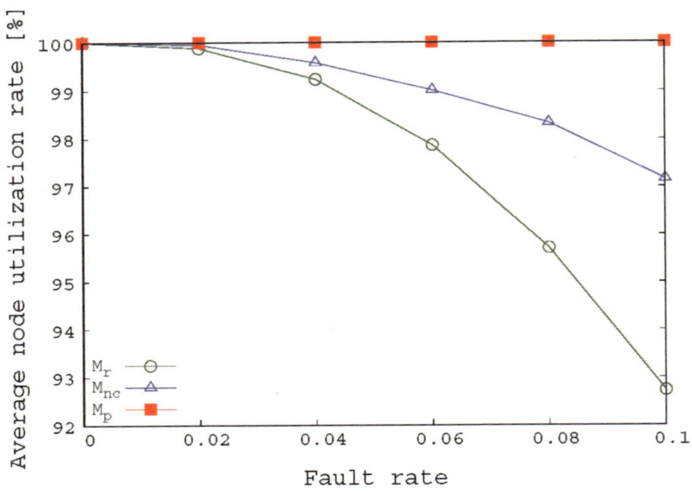

Figure 15.
Average node utilization rate vs. fault rate.

		f				
M_a	M_b	2%	4%	6%	8%	10%
M_r-1	M_{nc}-4	−94 (0.55)	−93 (0.45)	−92 (0.40)	−91 (0.35)	−89 (0.30)
M_p-1	M_{nc}-4	82 (0.75)	82 (0.60)	79 (0.50)	81 (0.45)	83 (0.40)

Table 2.
Maximum latency reduction rate $R(M_a, M_b)$ for the original algorithms.

values for any f. From this table, we found that M_p-1 reduces the average latency of M_{nc}-4 by about at least 79% without using additional VCs.

4.2.3 Performance comparison of routing algorithms with increased VCs

Next, the average latency of the three algorithms is compared by increasing the number of VCs twofold, threefold, and fourfold from the original (i.e. comparison of M_*-n, M_*-$2n$, M_*-$3n$, and M_*-$4n$). **Table 3** shows the results. The following can be found in the evaluation results:

1. With twofold VCs, the average latency can be reduced by about at least 66% and 33% for $f = 2$ and 10, respectively, compared with the original.

2. Fourfold increase in the number of VCs have only a marginal effect in reducing average latency.

3. Effect of latency reduction is higher in the algorithms with no VCs (i.e. M_r and M_p), and M_p shows the highest reduction rate for any f.

4.2.4 Performance comparison of routing algorithms with fixed number of VCs

Finally, the average latency of the three algorithms is compared under the same number of VCs (i.e. comparison of M_r-4, M_{nc}-4, and M_p-4). **Table 4** shows the maximum reduction rates of M_r-4 and M_p-4 for M_{nc}-4. By using four VCs, the

		f				
M_a	M_b	2%	4%	6%	8%	10%
M_r-1	M_r-2	67 (0.40)	63 (0.40)	55 (0.30)	55 (0.30)	55 (0.25)
	M_r-3	81 (0.45)	75 (0.40)	69 (0.35)	67 (0.30)	65 (0.30)
	M_r-4	84 (0.45)	78 (0.40)	73 (0.35)	71 (0.30)	69 (0.30)
M_{nc}-4	M_{nc}-8	66 (0.70)	54 (0.60)	40 (0.45)	35 (0.45)	33 (0.35)
	M_{nc}-12	73 (0.75)	60 (0.60)	47 (0.50)	38 (0.45)	41 (0.35)
	M_{nc}-16	76 (0.75)	62 (0.60)	49 (0.50)	40 (0.45)	44 (0.35)
M_p-1	M_p-2	88 (0.95)	82 (0.75)	78 (0.70)	76 (0.60)	75 (0.55)
	M_p-3	92 (1.00)	90 (0.80)	85 (0.70)	85 (0.65)	82 (0.55)
	M_p-4	92 (1.00)	91 (0.80)	89 (0.70)	87 (0.65)	86 (0.55)

Table 3.
Maximum latency reduction rate $R(M_a, M_b)$ for the algorithms with increased VCs.

		f				
M_a	M_b	2%	4%	6%	8%	10%
M_r-4	M_{nc}-4	−76 (0.60)	−75 (0.45)	−76 (0.40)	−75 (0.35)	−69 (0.35)
M_p-4	M_{nc}-4	96 (0.90)	96 (0.75)	94 (0.70)	94 (0.60)	94 (0.50)

Table 4.
Maximum latency reduction rate $R(M_a, M_b)$ for the algorithms with four VCs.

maximum reduction rates of M_r-4 and M_p-4 can be improved from the rates shown in **Table 2**, and M_p-4 always achieves more than 94% reduction rates for any f.

From the above results, we can conclude that, for reducing average latency of packet transmission, the reduction of hop count by the passage of faulty nodes, not always detour, is more effective than the avoidance of congestion using additional VCs.

4.3 Circuit amount

To evaluate the overhead of additional circuits such as switches, buffers, and links in the proposed approach, we designed two routers for M_r-1 and M_p-1 with Verilog HDL. In those routers, M_r and M_p are implemented into the routing circuits and the depth of input/output buffers is eight/one flits, respectively, as in the simulation setting. We used Xilinx Vivado EDA tool for synthesizing the routers for the target FPGA device of Vertex 7 xc7vx485tffg1761–2.

From the EDA tool, the router for M_r needs 1865 Look Up Tables (LUTs) in the FPGA device, while that for M_p needs 664 LUTs, which indicates about 64% LUT reduction. This is mainly because of the difference in the routing circuit; one routing circuit costs 193 LUTs for M_r and 18 LUTs for M_p. The small routing circuit is also benefit from the passage function. The additional circuits require only 27 LUTs, which is substantially small compared with the overall router circuit.

5. Conclusion and future work

We have introduced a novel approach for the design of fault-tolerant routing algorithms in 2D mesh NoCs. In contrast to the common idea of detouring faulty nodes, our approach allows passing through them with the slight modification of NoC architecture. We have provided a general methodology for designing fault-tolerant routing algorithms with the passage of faulty nodes, and as a design example, we have described the XY-based routing algorithm, showing how to prevent deadlocks in the routing rules. The XY-based routing algorithm allows passage of faulty nodes in the x-directional movement if the current and destination nodes are on the same row, while always allows in the y-directional movement.

To demonstrate the effect of the XY-based routing algorithm, we measured the average latency of packet transmission by computer simulations and compared with those of the well-known region-based algorithms proposed by Fukushima et al. and Chalasani et al. The results revealed that the XY-based algorithm reduced average latency of Chalasani's algorithm by about 79% without additional VCs and 94% with the same number of VCs. From the evaluation, we have found that passage is highly effective approach to reducing the average latency rather than employing VCs for congestion avoidance. We have also designed router circuit for the XY-based algorithm and showed that the overhead of additional circuit required for the proposed approach is substantially small compared with the overall router circuit.

As the passage of faulty nodes is a simple but effective approach, we have even more room to fully investigate the effect. For example, in this chapter, we selected popular XY routing as a base algorithm, which is a deterministic routing algorithm. Designing a new routing algorithm with the passage function based on some adaptive routing algorithm is a possible future research. As the passage is not limited to 2D mesh NoCs, designing passage-based fault-tolerant routing algorithms for other popular topology such as 2D torus, 3D mesh/torus is also one of the interesting future researches.

Acknowledgements

This work was supported by JSPS KAKENHI Grant Number JP18K11217.

Author details

Masaru Fukushi*† and Yota Kurokawa†
Graduate School of Sciences and Technology for Innovation, Yamaguchi University, Ube, Japan

*Address all correspondence to: mfukushi@yamaguchi-u.ac.jp

† These authors contributed equally.

References

[1] Mattson TG, Wijngaart RFV, Riepen M, Lehnig T, Brett P, Haas W, et al. The 48-core SCC Processor: the Programmer's View. In: Proceedings of International Conference for High Performance Computing, Networking, Storage and Analysis. New Orleans: IEEE; 13–19 Nov. 2010. p. 1–11.

[2] Sodani A, Gramunt R, Corbal J, Kim HS, Vinod K, Chinthamani S, et al. Knights landing: Second-generation intel xeon phi product. IEEE Micro. 2016;**36**(2):34-46. DOI: 10.1109/MM.2016.25

[3] Bohnenstiehl B, Stillmaker A, Pimentel JJ, Andreas T, Liu B, Tran AT, et al. KiloCore: A 32-nm 1000-processor computational array. IEEE Journal of Solid-State Circuits. 2017;**52**(4): 891-902. DOI: 10.1109/JSSC.2016.2638459

[4] Dally WJ, Towles B. Route packets, not wires: on-chip interconnection networks. In: Proceedings of the 38th Design Automation Conference. Las Vegas: IEEE; 22 June 2001; p. 684–689. DOI: 10.1109/DAC.2001.156225

[5] Dally WJ, Towles B. Principles and practices of interconnection networks. Morgan Kaufman Publishers; 2004.

[6] Hsin HK, Chang EJ, Lin CA, Wu AY. Ant colony optimization-based fault-aware routing in mesh-based network-on-chip systems. IEEE Transactions on Computer-Aided Design of Integrated Circuits and Systems. 2014;**33**(11): 1693-1705. DOI: 10.1109/TCAD.2014.2347922

[7] Liu J, Harkin J, Li Y, Maguire LP. Fault-tolerant networks-on-chip routing with coarse and fine-grained look-ahead. IEEE Transactions on Computer-Aided Design of Integrated Circuits and Systems. 2016;**35**(2):260-273. DOI: 10.1109/TCAD.2015.2459050

[8] Zhao H, Bagherzadeh N, Wu J. A general fault-tolerant minimal routing for mesh architectures. IEEE Transactions on Computers. 2017;**66**(7): 1240-1246. DOI: 10.1109/TC.2017.2651828

[9] Janfaza V, Baharlouei E. A new fault-tolerant deadlock-free fully adaptive routing in noc. In: Proceedings of IEEE East-West Design & Test Symposium (EWDTS); 29 Sept.-2 Oct. 2017. Novi Sad: Serbia. IEEE; 2017. p. 1-6. DOI: 10.1109/EWDTS.2017.8110139

[10] Sinha D, Roy A, Kumar KV, Kulkarni P, Soumya J. Dn-FTR: Fault-tolerant routing algorithm for mesh based network-on-chip. In: Proceedings of the 4th International Conference on Recent Advances in Information Technology (RAIT); 15–17 March 2018; Dhanbad, India. IEEE; 2018. p. 1–5 DOI: 10.1109/RAIT.2018.8389083

[11] Wang L, Wang X, Wang Y. An approximate bufferless network-on-chip. IEEE Access. 2019;7:141516–141532. DOI: 10.1109/ACCESS.2019.2943922

[12] Chen KH, Chiu GM. Fault-tolerant routing algorithm for meshes without using virtual channels. Journal of Information Science and Engineering. 1998;**14**(4):765-783

[13] Holsmark R, Kumar S. Corrections to chen and chiu's fault tolerant routing algorithm for mesh networks. Journal of Information Science and Engineering. 2007;**23**(6):1649-1662

[14] Fu B, Han Y, Li H, Li X. ZoneDefense: A fault-tolerant routing for 2-D meshes without virtual channels. IEEE Transactions on Very Large Scale Integration Systems. 2014; **22**(1):113-126. DOI: 10.1109/TVLSI.2012.2235188

[15] Fukushima Y, Fukushi M, Yairi IE. A region-based fault-tolerant routing algorithm for 2D irregular mesh network-on-chip. Journal of Electronic Testing. 2013;**29**:415-429. DOI: https://doi.org/10.1007/s10836-013-5377-9

[16] Wu J. A fault-tolerant and deadlock-free routing protocol in 2D meshes based on odd-even turn model. IEEE Transactions on Computers. 2003;**52**(9): 1154-1169. DOI: 10.1109/TC.2003.1228511

[17] Chalasani S, Boppana RV. Communication in multicomputers with nonconvex faults. IEEE Transactions on Computers. 1997;**46**(5):616-622. DOI: 10.1109/12.589238

[18] Kurokawa Y, Fukushi M. Passage of faulty nodes: A novel approach for fault-tolerant routing on nocs. IEICE Transactions on Fundamentals of Electronics, Communications and Computer Sciences. 2019;**E102-A**(12): 1702-1710. DOI: 10.1587/transfun.E102.A.1702

[19] Takanami I, Horita T, Akiba M, Terauchi M, Kanno T. A built-in self-repair circuit for restructuring mesh-connected processor arrays by direct spare replacement. Transactions on Computational Science XXVII. 2016; **9570**:97-119. DOI: https://doi.org/10.1007/978-3-662-50412-3_7

Chapter 5

Digital Control of Active Network Microstructures on Silicon Wafers

Zhongjing Ren, Jianping Yuan and Peng Yan

Abstract

This chapter presents a promising digital control of active microstructures developed and tested on silicon chips by current division and thus independent Joule heating powers, especially for planar submillimeter two-dimensional (2-D) grid microstructures built on silicon wafers by surface microfabrication. Current division on such 2-D grid networks with 2 × 2, 3 × 3, and n × n loops was modeled and analyzed theoretically by employing Kirchhoff's voltage law (KVL) and Kirchhoff's current law (KCL), which demonstrated the feasibility of active control of the networks by Joule heating effect. Furthermore, *in situ* testing of a typical 2-D microstructure with 2 × 2 loops by different DC sources was carried out, and the thermomechanical deformation due to Joule heating was recorded. As a result, active control of the current division has been proven to be a reliable and efficient approach to achieving the digital actuation of 2-D microstructures on silicon chips. Digital control of such microstructural networks on silicon chips envisions great potential applications in active reconfigurable buses for microrobots and flexible electronics.

Keywords: surface microfabrication, current division, joule heating, digital control, grid microstructures

1. Introduction

Silicon-based microelectromechanical systems (MEMS) devices, including sensors, actuators, and generators, show wide applications in microrobots [1], medical devices [2, 3], and flexible electronics [4]. Such miniaturized systems, on one hand, offer great potentials for improving the abilities of micromanipulation and functioning in some extreme conditions, such as limited working space and large displacements; On the other hand, however, the strong requirement of precise and effective control of these kinds of devices is not easy. Therefore, microstructures allowing for reliable and efficient actuation and large displacement are worth investigating.

Thermal microactuators have been proven to be able to realize large displacements more efficiently, and a variety of materials, for example, ceramics [5–7], polymers [8], composites [1, 9, 10], and metals [11–14], are available for building such active microstructures. Moreover, the geometries of these microstructures heavily depend on the selected materials, among which electrothermal bilayer beams have obvious advantages on large displacements, low costs, and high compatibility with mature microfabrication processes. Many efforts have been paid to develop electrothermal microstructures based on different applications and requirements. A safety and arming device composed of two V-shape electrothermal actuators was built, and the design of a cascaded V-beam amplification and two mechanical sliders enabled large deformation. As a result, a large planar displacement of 231.78 μm was achieved by applying a voltage of 15 V [15]. A typical U-shape electrothermal actuator made of a single material allowing for planar bending due to the thermal mismatch between the cold and hot arms was developed. Since the Joule heating power on the narrower arm was smaller than the wider one, the thermal expansion on the narrower (or hot) arm is larger than that of the wider (or cold) arm [16]. Another representative study is related to high frequency, low power, electrothermal bimorph actuators with shape memory effects, and the significant thermal mismatch and shape memory effect contributed to very large out-of-plane deformation [17].

However, previous research on electrothermal actuators usually focused on development beams with simple geometries, such as bridges, V-shape, U-shape, etc. The current flows through the beams were the same, which limited their ability of reconfiguration. To accomplish diverse reconfiguration of such electrothermal microstructures, active current division across planar (or 2-D) bilayer microstructures offers a promising approach to digital control of microstructures for distributed thermal balance and thus various deformations.

Our group has been endeavoring to design, fabrication, and characterization of 2-D multilayered microstructure consisting of beams for out-of-plane deflection, vertical deployment [18–24], and twisting under electrothermal actuation [9]. These microstructures can be created on whole wafers and tested separately by cutting the silicon wafer into chip-scale pieces [18]. Active parts of such microstructures are released by selective etching on the silicon substrates, while the anchored parts are protected from being etched. Different Joule heating powers and balanced temperatures on the grid networks could be obtained. However, instead of qualitative analysis, it is more important and meaningful to quantify the effect of electrothermal effect of bilayer 2-D microstructures, which will lay a solid foundation for modeling and predicting their potentials for large out-of-plane displacement.

The rest of this chapter is organized as follows. In Section 2, theoretical analysis of current divisions in some typical 2-D grid microstructures is firstly carried out, followed by the quantitative analysis of the equivalent resistance and Joule heating power. After that, fabrication and characterization of electrothermal and thermomechanical performances are presented experimentally in Section 3. The results are then shown and discussed in Section 4. Finally, the chapter is concluded in Section 5.

2. Theoretical analysis of current divisions

In this section, Kirchhoff's Voltage Law (KVL) and Kirchhoff's Current Law (KCL) are employed appropriately to figure out the feasible current division in several representative 2-D grid structures. Note that the scale of the 2-D structures does not change the relative division ratios between these beams. So, we consider more general structures, instead of only microstructures, in this section. It also hints that such cross-scale research on the current distribution is applicable to multiple surroundings and uses. For the sake of simplicity, all the grid structures presented in this chapter are composed of bilayer beams with the same materials and dimensions. Furthermore, these beams are incorporated into grid networks with different geometries. Two representative 2-D grid structures with the 2×2 loops and 3×3 loops are shown in **Figure 1**, where the candidate input ports are marked as red dots. The 2×2 grid structure consists of 12 beams for the current division, while the

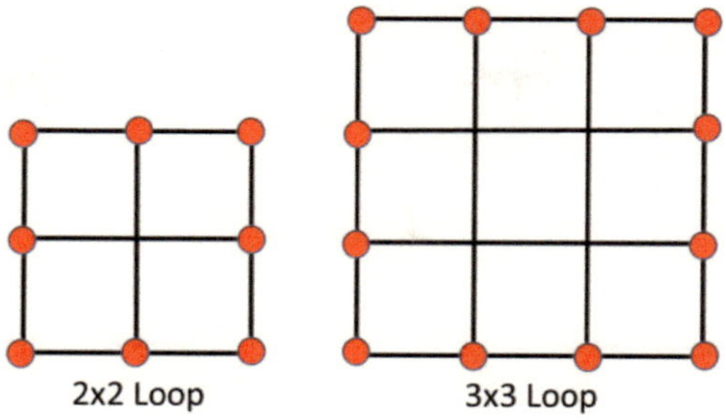

Figure 1.
Schematic views of two planar grid structures for the current division.

3 × 3 one consists of 24 beams. Besides that, it is worth noting that grid structures with n × n loops can be similarly proven to be made of 2n(n + 1) beams.

Let us start with studying the current division through 2 × 2 grid structures by assuming that a constant voltage *V* is applied to any two outer red nodes, as seen in **Figure 1**. To ensure the stable connection between the voltage source and the red input nodes, another two supporting beams with fixed ends anchored on silicon wafers are designed. Hence, it can be derived simply that there are six independent cases in total for voltage inputs, as illustrated in **Figure 2**. The resistances of the beams in grid structures, as well as those of the two supporting beams, are assumed to be the same *R*. To acquire the current distribution through these beams of all the cases, KCL and KVL are adopted to the nodes and loops, respectively. Taking Case 1 as an example, as presented in **Figure 3**, nine independent KCL equations at the nodes N_i (i = 1, 2, ... ,9) and five independent KVL ones at the loops C_i (i = 1,2, ... ,5) are established, as shown in Eq. [1].

Specifically, the KCL equations could be written as

$$\begin{cases} I_1 - I_2 - I_4 = 0 \\ I_2 - I_3 - I_5 = 0 \\ I_3 - I_6 = 0 \\ I_4 - I_7 + I_8 + I_{10} = 0 \\ I_5 - I_8 + I_9 - I_{11} = 0 \\ I_6 - I_9 - I_{12} = 0 \\ -I_{10} + I_{13} = 0 \\ I_{11} - I_{13} + I_{14} = 0 \\ I_{12} - I_{14} = 0 \end{cases} \tag{1}$$

while the KVL equations are formulated as

$$\begin{cases} V - (I_1 + I_4 + I_7) \cdot R = 0 \\ I_4 \cdot R - (I_2 + I_5 + I_8) \cdot R = 0 \\ I_5 \cdot R - (I_3 + I_6 + I_9) \cdot R = 0 \\ I_8 \cdot R - (I_{10} + I_{11} + I_{13}) \cdot R = 0 \\ (I_9 + I_{11}) \cdot R - (I_{12} + I_{14}) \cdot R = 0 \end{cases} \tag{2}$$

As a result, the 14 unknowns I_i (i = 1, 2, ... , 14) can be uniquely solved by the derived 14 independent equations, that is,

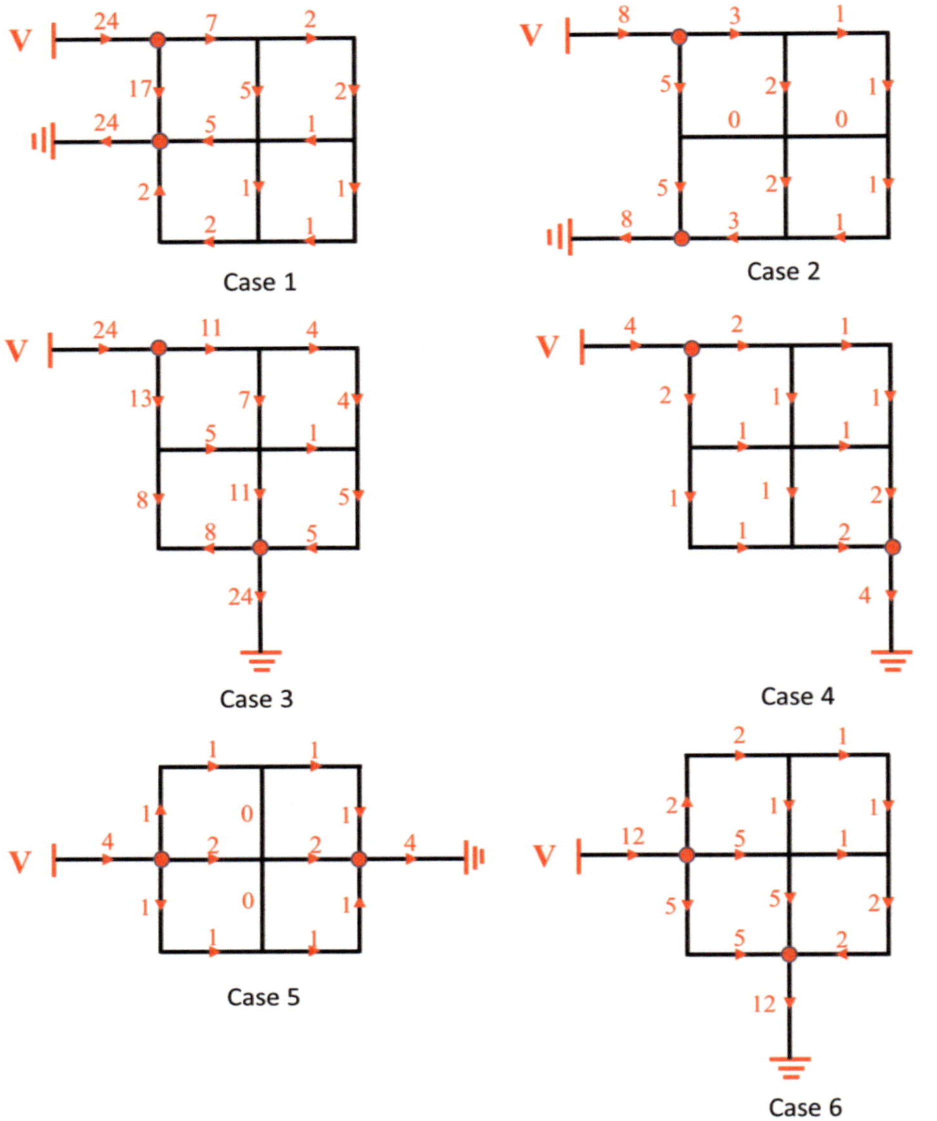

Figure 2.
Six independent cases of voltage inputs into grid structures with 2 × 2 loops. The red arrows denote the current directions, while the red numbers represent the current division factors between different beams in each case. Note that factors of beams without current flow are 0.

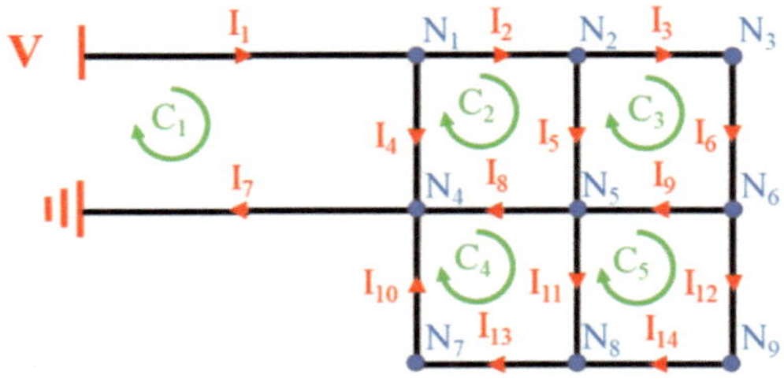

Figure 3.
KCL and KVL on case 1 of the grid structures with 2 × 2 loops and two supporting beams.

$$\begin{cases} I_1 = I_7 = \dfrac{24}{65} \cdot \dfrac{V}{R} \\ I_2 = \dfrac{7}{65} \cdot \dfrac{V}{R} \\ I_3 = I_6 = I_{10} = I_{13} = \dfrac{2}{65} \cdot \dfrac{V}{R} \\ I_4 = \dfrac{17}{65} \cdot \dfrac{V}{R} \\ I_5 = I_8 = \dfrac{5}{65} \cdot \dfrac{V}{R} \\ I_9 = I_{11} = I_{12} = I_{14} = \dfrac{1}{65} \cdot \dfrac{V}{R} \end{cases} \tag{3}$$

which demonstrates the determination of current division factors for all the beams in Case 1. Similarly, the current division factors in Cases 2–6 can be obtained uniquely, as seen in **Figure 2**.

Further investigation on the current distribution across planar structures with 3×3 loops was taken, and the division factors and directions of the current flows had been proved to be unique. Generally, there are 12 independent cases for voltage inputs, as seen in **Figure 4**. Particularly, current distributions of two representative geometries with 3×3 loops, Case 3 and Case 6, are solved and illustrated in **Figure 5**. A very interesting phenomenon observed in Case 3 of the 3×3 loops structure is that the current division factors are symmetric about the central axis marked as a green dashed line in **Figure 5**. Further exploration reveals that the

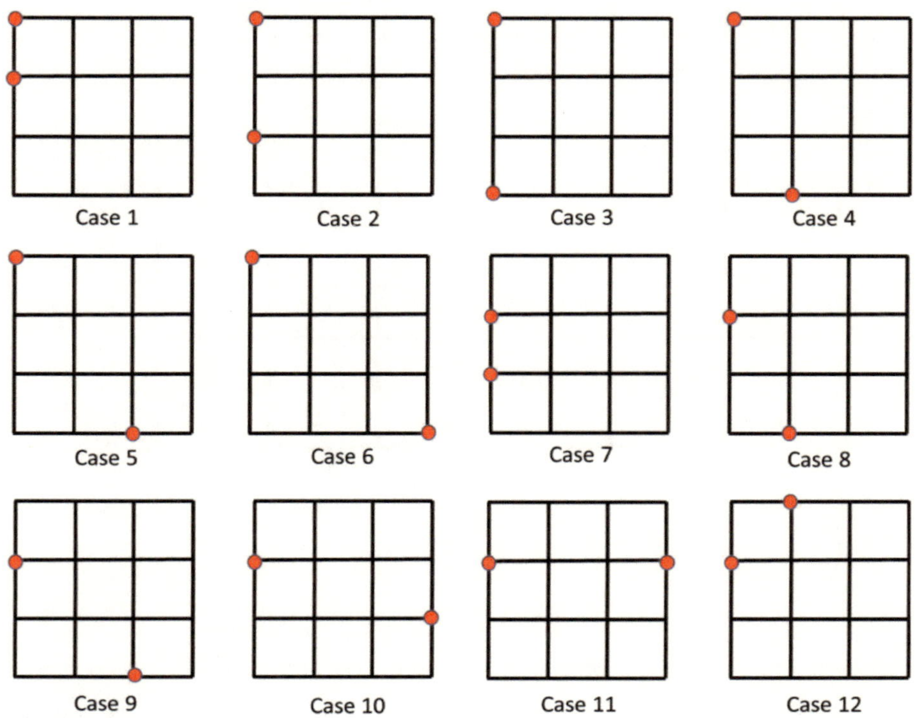

Figure 4.
Twelve independent cases of voltage inputs into grid structures with 3×3 loops.

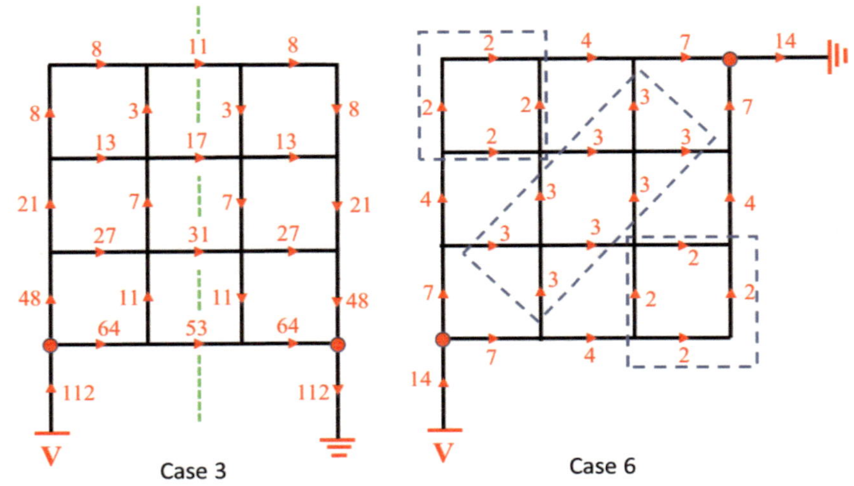

Figure 5.
Current distributions of two representative cases of voltage inputs into grid structures with 3 × 3 loops.

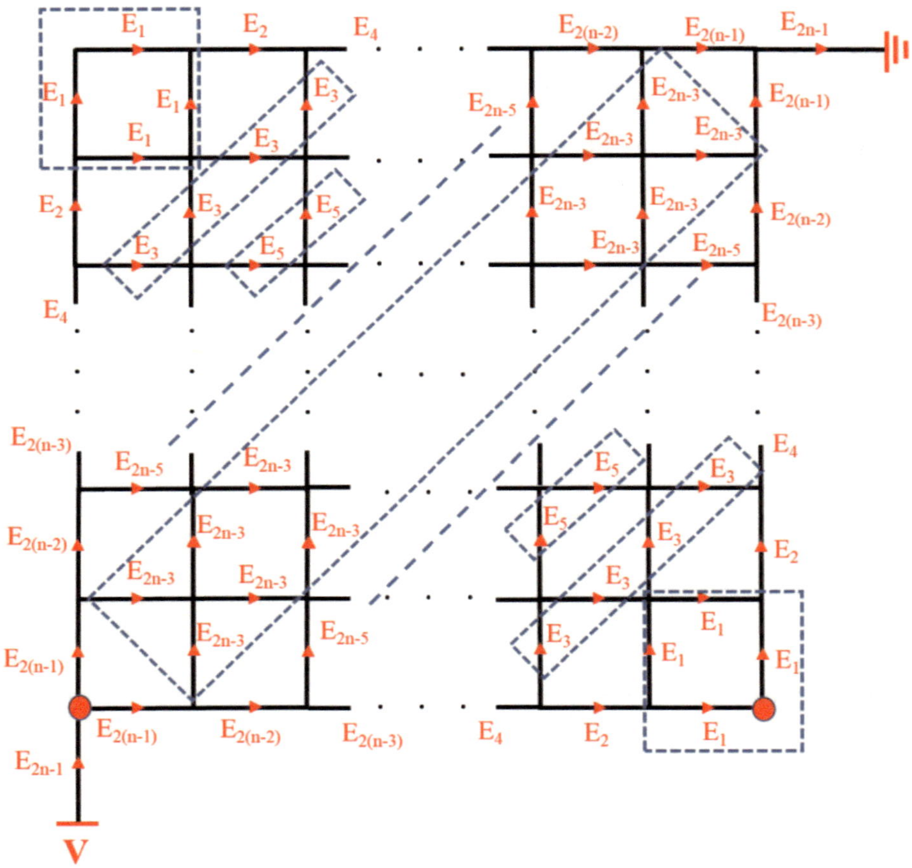

Figure 6.
*Current distribution across an n × n loops geometry like Case 6 in **Figure 5**.*

directions of current flows symmetrical about the green line are opposite, and the magnitudes of the currents closer to the input ports tend to be larger. In addition, though the distribution regularity of Case 6 is a little more complicated than that of

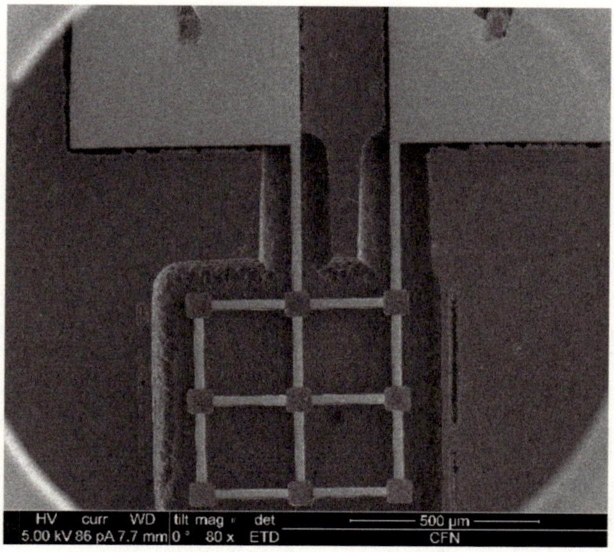

Figure 7.
SEM image of a representative 2-D microstructure consisting of bilayer beams.

the Case 3, a very interesting symmetric current distribution about the diagonal of the geometry is found. Similarly, current flows through more complex geometries with n × n loops like Case 6 are shown in **Figures 6** and **7**. It is worth noting that the corresponding current division factors in **Figure 6** are represented by E_i ($E_2 = 2E_1$, $E_n = E_{n-1} + E_{n-2}$ when $n \geq 3$ and n is odd, $E_n = (E_{n-1} + E_{n-2})/2$ when $n \geq 3$ and n is even). Generally, there are (2n − 2) different currents through the n × n grid network, enabling digital control of diverse and regional Joule heating powers across such a 2-D network.

Based on the current divisions of 2 × 2 loops structures with diverse voltage inputs' cases presented and discussed in this section, the effect of Joule heating on such conductive grid structures is available to be quantified and evaluated.

Assuming that the resistances of a single beam in the 2 × 2 loops, as well as each of the supporting beams, are the same, R, and the external voltages applied are V, The equivalent resistances of each case of the 2 × 2 grid structures shown in **Figure 2** can be solved using KCL and KVL laws, which was presented in the previous research. The Joule heating power of the grid structures and supporting beams can be then determined and listed in **Table 1**, where R_i represents the resistance in case i, while the P_i represents the heating power when a voltage of V applied. It can be seen clearly from **Table 1** that equivalent resistances are approximately three times over the resistance of a single beam, and thus the expected Joule heating power is about one-third of the power when a single beam is applied by a voltage of V.

Similarly, the equivalent resistances of the 3 × 3 loops structure in Case 3 and Case 6 can be derived to be 293/56·R and 27/7·R, respectively.

Case (i)	1	2	3	4	5	6
Equivalent Resistance (R)	65/24	13/4	77/24	7/2	3	17/6
Heating Power (V^2/R)	24/65	4/13	24/77	2/7	1/3	6/17

Table 1.
Equivalent resistances and joule heating powers of the six cases of 2 × 2 loops with two supporting beams.

3. Experimental validation of 2-D microstructures

3.1 Fabrication of 2-D grid microstructures

To demonstrate the effect of electrothermal actuation of the 2-D structures, a series of ultrathin (or 2-D) microstructures consisting of grid beams and supporting beams that are mentioned in Section 2 are designed, fabricated, and tested.

Several previous research by our group can be referred to on a typical design and fabrication processes of such 2-D microstructures with two different materials. Specifically, aluminum and NiTi alloys (which are in the austenite phase in the range of testing temperatures) are selected as the bottom layer and top layer, respectively. What needs to be emphasized here is that although NiTi alloys show great potential for shape memory effects, this effect is not introduced intentionally in this research. It is the effect of digital Joule heating that we want to present in this chapter. Aluminum was chosen due to its significantly larger coefficient of thermal expansion than that of the NiTi alloys in the austenite phase.

As a result, a typical 2-D microstructure with the geometry presented in **Figure 3** was imaged by the SEM after being selectively released from the silicon chip, as seen in **Figure 7**. It is worth noting that these two contact pads attached to the silicon chip were connected to gold wires with a diameter of 20 μm. The gold wire was used for electrical signal transfer from the logic printed circuit board after being fixed on the *in situ* imaging stage in SEM.

3.2 Results and discussion of *in situ* test of the microstructure

The experimental setup for *in situ* electrothermal actuation of the microstructure is illustrated in **Figure 8**. The stage of the SEM was tilted from zero degree to 45 degree for easier observation and measurement, and the configuration of the microstructure without heating was reimaged as shown in **Figure 9**. The *in situ* electrothermal testing of such a microstructure on the silicon chip started with the

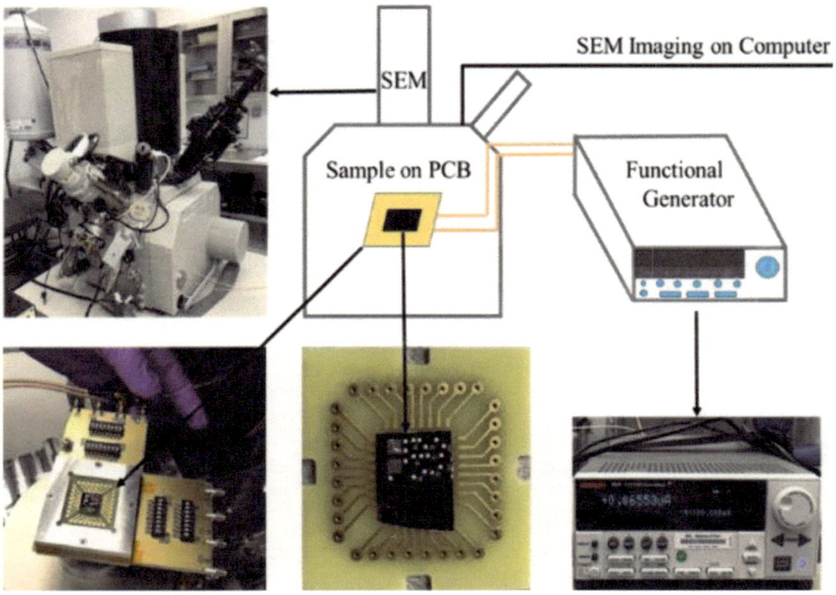

Figure 8.
Experimental setup of microstructures on the silicon chip for testing.

Figure 9.
SEM image of the microstructure after being tilted by 45 degrees.

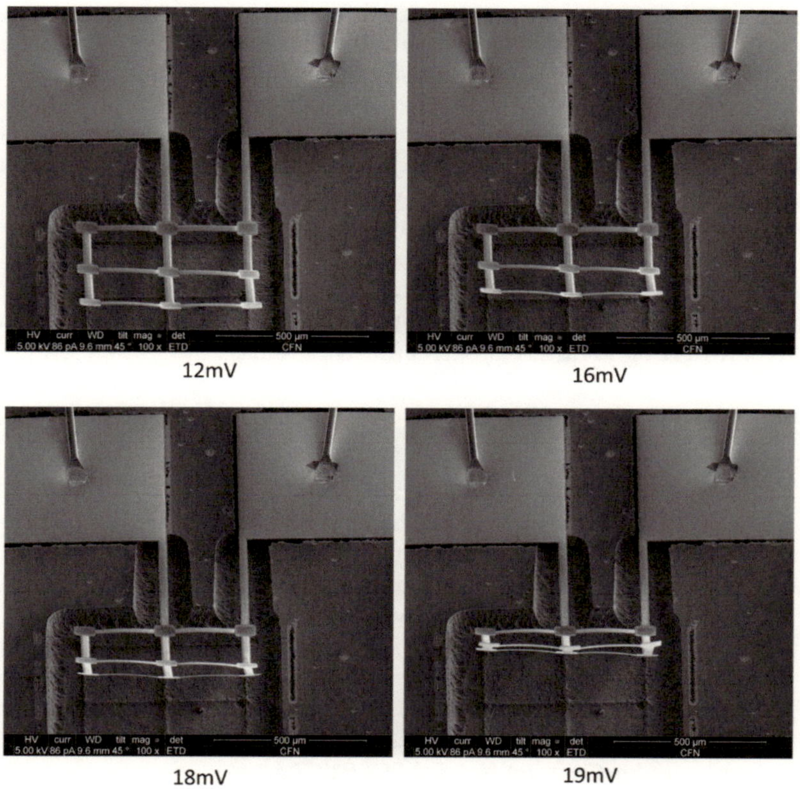

Figure 10.
SEM images of microstructures under different driving voltages in electrothermal testing.

application of a DC source by the Agilent 4155C Semiconductor Parameter Analyzer. The shapes of the microstructure under constant voltages of 12 mV, 15 mV, 18 mV, and 19 mV were imaged and presented in **Figure 10**, and a supplementary video about this process was recorded simultaneously. It can be seen from **Figure 10**, as well as the video, that the microstructure could be digitally actuated by distributed currents for diverse and regional Joule heating. Therefore, the effect of active control of microstructures using digitally distributed currents is demonstrated. It is important to highlight that although the "digital currents" were inherently ensured, the thermomechanical reconfiguration of corresponding beams does not seem to be that "digital". It could be explained by the effect of scaling which could have a significant influence on the thermal conduction between the beams. The scaling effect is expected to be alleviated gradually with structural scale-up. Generally, such silicon chip-based microfabrication processes show great compatibility, effectivity, and efficiency in the development and validation of 2-D microstructures.

4. Conclusions

In conclusion, effective and efficient development of active control of 2-D microstructures based on silicon chips is presented in this chapter. Representative planar structures composed of grid beams are introduced to quantitatively analyze possible current distribution across the conductive geometry using KCL and KVL. Diverse current divisions of structures with different loops and voltage inputs have been proven to be available for digital control of electrothermal actuators. Besides that, the determination of equivalent resistances and resulting Joule heating powers have contributed to the evaluation of *in situ* experiments on the representative microstructure created on silicon chips by microfabrication. The process and critical steps of thermomechanical deformation of such a microstructure are shown to demonstrate the effect of digital control by Joule heating. The unsignificant digital deformation could be attributed to the scale effect of thermal conduction.

It is worth highlighting that much more various and precise current divisions can be obtained by superposition of different voltage inputs, which can be an attractive topic in the future. Another promising research is to investigate the scale effect on Joule heating in different current distributions.

Acknowledgements

The authors would like to thank Dr. Chang-Yong Nam, Dr. Camino Fernando, and Dr. Ming Lu from the Center of Functional Nanomaterials, Brookhaven National Laboratory. In addition, I really appreciate the great supports and suggestions from Robert Bauer and Yang Xu from Stevens Institute of Technology. The National Natural Science Foundation of China (No.11572248) and China Scholarship Council has in part supported the research. The research was in part carried out at the Center for Functional Nanomaterials (CFN), Brookhaven National Laboratory (BNL), which is supported by the U.S. Department of Energy, Office of Basic Energy Sciences, under Contract No. DE-SC0012704.

Conflict of interest

The authors declare no conflict of interest.

Author details

Zhongjing Ren[1]*, Jianping Yuan[2] and Peng Yan[1]

1 Key Laboratory of High-Efficiency and Clean Mechanical Manufacture (Ministry of Education), School of Mechanical Engineering, Shandong University, Jinan, Shandong Province, China

2 National Key Laboratory of Space Flight Dynamics, School of Astronautics, Northwestern Polytechnical University, Xi'an, Shaanxi Province, China

*Address all correspondence to: zren@sdu.edu.cn

References

[1] Su X, Ren Z, Yan J, Shi Y, Pan Q. Microstructure and twisting ability of an adjusted antisymmetric angle ply laminate. Applied Physics Letters. 2019; **114**(21):211902. DOI: 10.1063/1.5089809

[2] Chun YJ, Levi DS, Mohanchandra KP, Fishbein MC, Carman GP. Novel micro-patterning processes for thin film NiTi vascular devices. Smart Materials and Structures. 2010;**19**(10):105021. DOI: 10.1088/0964-1726/19/10/105021

[3] Ghazali FAM, Hasan MN, Rehman T, Nafea M, Ali MSM, Takahata K. Micro-electromechanical-system actuators for biomedical applications: A review. Journal of Micromechanics and Microengineering. 2020;**30**(7):073001. DOI: 10.1088/1361-6439/ab8832

[4] Kamyshny A, Magdassi S. Conductive nanomaterials for 2D and 3D printed flexible electronics. Chemical Society Reviews. 2019;**48**(6): 1712-1740. DOI: 10.1039/C8CS00738A

[5] Galos R, Shi Y, Ren Z, Zhou L, Sun H, Su X, et al. Electrical impedance measurements of PZT nanofiber sensors. Journal of Nanomaterials. 2017; **2017**(Special Issue):8275139. DOI: 10.1115/DETC2016-59687

[6] Galos R, Shi Y, Ren Z, Synowicki R, Sun H, Nykypanchuk D, et al. The dielectric constant of PZT nanofiber at visible and NIR wavelengths. Nano-Structures and Nano-Objects. 2018;**15**: 205-211. DOI: 10.1016/j.nanoso.2017.10.001

[7] Galos R, Shi Y, Ren Z, Sun H. Electrical impedance matching of PZT nanogenerators. In: International Design Engineering Technical Conferences and Computers and Information in Engineering Conference; 6–9 August 2017; Cleveland, Ohio, USA. p. 47 V004T09A025

[8] Al-Zandi MH, Wang C, Voicu R, Muller R. Measurement and characterisation of displacement and temperature of polymer based electrothermal microgrippers. Microsystem Technologies. 2018;**24**(1): 379-387. DOI: 10.1007/s00542-017-3298-8

[9] Su X, Ren Z, Sun H, Shi Y, Pan Q. The submicron fabrication process for T gate with a flat head. In: International Design Engineering Technical Conferences and Computers and Information in Engineering Conference; 26–29 August 2018; Quebec City, Quebec, Canada. p. V004T08A018

[10] Su X, Ren Z, Pan Q, Lu M, Camino F, Shi Y. Design, modeling and experimental validation of a micro cantilever beam with an electro-controllable twisting ability. Journal of Micromechanics and Microengineering 2021;**31**(6):065010. DIO: 10.1088/1361-6439/abfc35

[11] Sun H, Luo J, Ren Z, Lu M, Shi Y. Effects of deposition and annealing conditions on the crystallization of NiTi thin films by e-beam evaporation. Micro and Nano Letters. 2020;**15**(10):670-673. DIO: 10.1049/mnl.2020.0004

[12] Sun H, Luo J, Ren Z, Lu M, Nykypanchuk D, Mangla S, et al. Shape memory alloy bimorph microactuators by lift-off process. Journal of Micro and Nano-Manufacturing. 2020;**8**(3): 031003. DOI: 10.1115/1.4048146

[13] Ren Z, Yuan J, Su X, Sun H, Galos R, Shi Y. A new fabrication process for microstructures with high area-to-mass ratios by stiffness enhancement. In: International Design Engineering Technical Conferences and Computers and Information in Engineering Conference; 26–29 August 2018; Quebec City, Quebec, Canada. p. V004T08A034

[14] Ren Z, Yuan J, Su X, Sun H, Galos R, Shi Y, et al. Vertical deployment of multilayered metallic microstructures with high area-to-mass ratios by thermal actuation. Journal of Micro and Nano-Manufacturing. 2019;7(3):031002. DOI: 10.1115/1.4043987

[15] Li X, Zhao Y, Hu T, Xu W, Zhao Y, Bai Y, et al. Design of a large displacement thermal actuator with a cascaded V-beam amplification for MEMS safety-and-arming devices. Microsystem Technologies. 2015;**21**(11):2367-2374. DOI: 10.1007/s00542-015-2447-1

[16] Hussein H, Tahhan A, Le Moal P, Bourbon G, Haddab Y, Lutz P. Dynamic electro-thermo-mechanical modelling of a U-shaped electro-thermal actuator. Journal of Micromechanics and Microengineering 2016;**26**(2):025010. DOI: 10.1088/0960-1317/26/2/025010

[17] Knick CR, Sharar DJ, Wilson AA, Smith GL, Morris CJ, Bruck HA. High frequency, low power, electrically actuated shape memory alloy MEMS bimorph thermal actuators. Journal of Micromechanics and Microengineering 2019;**29**(7):075005. DOI: 10.1088/1361-6439/ab1633

[18] Ren Z. Design, fabrication, and control of reconfigurable active microstructures for solar sails [thesis]. Stevens Institute of Technology; 2020

[19] Ren Z, Yuan J, Shi Y. Electro-thermo-mechanical modelling of micro solar sails of chip scale spacecraft in space. Microsystem Technologies. 2021;**27**(12):4209-4215. DOI: 10.1007/s00542-020-05204-x

[20] Ren Z, Yuan J, Su X, Bauer R, Xu Y, Mangla S, et al. Current divisions and distributed Joule heating of two-dimensional grid microstructures. Microsystem Technologies. 2021;**27**(9): 3339-3347. DOI: 10.1007/s00542-020-05103-1

[21] Ren Z, Yuan J, Su X, Mangla S, Nam C-Y, Lu M, et al. Thermo-mechanical modeling and experimental validation for multilayered metallic microstructures. Microsystem Technologies. 2021;**27**(7):2579-2587. DOI: 10.1007/s00542-020-04988-2

[22] Ren Z, Yuan J, Su X, Mangla S, Nam C-Y, Lu M, et al. Electro-thermal modeling and experimental validation for multilayered metallic microstructures. Microsystem Technologies. 2021;**27**(5):2041-2048. DOI: 10.1007/s00542-020-04964-w

[23] Ren Z, Yuan J, Su X, Shi Y. A novel design and thermal analysis of micro solar sails for solar sailing with chip scale spacecraft. Microsystem Technologies. 2021;**27**(7):2615-2622. DOI: 10.1007/s00542-020-05094-z

[24] Ren Z, Yuan J, Su X, Xu Y, Bauer R, Mangla S, et al. Multilayered microstructures with shape memory effects for vertical deployment. Microsystem Technologies. 2021;**27**(9): 3325-3332. DOI: 10.1007/s00542-020-05101-3